3rd International Conference on X-ray Technique 2016

Journal of Physics: Conference Series Volume 808

St. Petersburg, Russia
24 – 25 November 2016

Editors:

N.N. Potrakhov
A. Yu. Gryaznov
A.A. Lisenkov
D.K. Kostrin

ISBN: 978-1-5108-3682-2
ISSN: 1742-6588

Printed from e-media with permission by:

Curran Associates, Inc.
57 Morehouse Lane
Red Hook, NY 12571

Some format issues inherent in the e-media version may also appear in this print version.

Printed by Curran Associates, Inc. (2017)

For permission requests, please contact the Institute of Physics
at the address below.

Institute of Physics
Dirac House, Temple Back
Bristol BS1 6BE UK

Phone: 44 1 17 929 7481
Fax: 44 1 17 920 0979

techtracking@iop.org

Additional copies of this publication are available from:

Curran Associates, Inc.
57 Morehouse Lane
Red Hook, NY 12571 USA
Phone: 845-758-0400
Fax: 845-758-2633
Email: curran@proceedings.com
Web: www.proceedings.com

Table of contents

Volume 808

3rd International Conference on X-ray Technique
24–25 November 2016, St Petersburg, Russian Federation

Accepted papers received: 7 February 2017
Published online: 21 February 2017

Preface

Papers

Past, Present and Future of the X-ray Technique – a View from the 3rd International Conference on X-ray Technique

N N Potrakhov[1], **A Yu Gryaznov**[1], **A A Lisenkov**[2] **and D K Kostrin**[1]

[1] Saint Petersburg Electrotechnical University “LETI”, 197376, Saint Petersburg, Russia
[2] Institute of Problems of Mechanical Engineering, Russian Academy of Sciences, 199178, Saint Petersburg, Russia

E-mail: kzhamova@gmail.com

Abstract. In this preface a brief history, modern aspects and future tendencies in development of the X-ray technique as seen from the 3rd International Conference on X-ray Technique that was held on 24–25 November 2016 in Saint Petersburg, Russia are described

On 24–25 November 2016 in Saint Petersburg on the basis of Saint Petersburg State Electrotechnical University "LETI" n. a. V. I. Ulyanov (Lenin) was held the 3rd International Conference on X-ray Technique. The tradition to hold a similar conference in our country was laid in Soviet times. The last of them, the All-Union Conference on the Prospects of X-ray Tubes and Equipment was organized and held more than a quarter century ago – on 21–23 November 1999, at the initiative and under the leadership of the chief engineer of the Leningrad association of electronic industry "Svetlana" Borovsky Alexander Ivanovich and the chief of special design bureau of X-ray devices of "Svetlana" Shchukin Gennady Anatolievich. The most active part in the organization and work of the conference played members of the department of X-ray and electron beam instruments of Leningrad Electrotechnical Institute "LETI" (the former name of Saint Petersburg State Electrotechnical University "LETI"), represented by head of the department professor Ivanov Stanislav Alekseevich.

Choice of the special design bureau of X-ray devices as the site for the Conference at that time was not accidental. Total production of X-ray tubes of all types in the mid 80-ies of the 20th century amounted to tens of thousands of units per year. The company was the leading and almost the only manufacturer of X-ray tubes; had a unique experience and the most modern at that time vacuum equipment and technologies.

However, as a result of the collapse of the Soviet Union and the subsequent economic crisis, domestic industry of X-ray apparatuses production was largely destroyed and had long been in decline. Were broken once close scientific and technical relationships between developers, manufacturers and users of X-ray equipment, many of the technologies were lost and, most importantly, the experienced hardware manufacturers – the industry's elite – went out of business.

In recent years, in connection with the necessity of meeting the country's needs in X-ray equipment were revived some old and founded new enterprises of this industry. Practice shows that many of them compete successfully in certain areas of X-ray apparatuses production with leading international manufacturers. It is time to assess the state of affairs in this industry, in particular, the level of scientific and technological development, and also try to identify problem areas and possible sources for further development. In this regard, at the initiative of the leading lecturers of the department of electronic instruments and devices of Saint Petersburg State Electrotechnical University "LETI" was revived the Conference for researchers and producers of X-ray technique in our country.

Figure 1. Opening of the Conference.

The venue and the organizers of this post-Soviet Conference were determined not by chance: JSC "Svetlana-X-Ray" is still the largest developer and manufacturer of X-ray tubes not only in our country but in the world as well and Saint Petersburg State Electrotechnical University "LETI", in the face of the department of electronic instruments and devices – the current name of the department of X-ray and electron beam instruments, is the country's only higher educational institution training specialists directly in the development of X-ray tubes.

It is officially considered that the department of electronic instruments and devices was founded in October 1938. It was created on the initiative of professor Alexander Aleksandrovich Shaposhnikov – the founder of domestic scientific school on vacuum electronics. As head of the department was appointed Feofan Nikolaevich Kharadzha – a graduate of Leningrad Electrotechnical Institute "LETI", who have taken the long path from the engineer to the head of the laboratory of X-ray tubes in JSC "Svetlana". The first in our country X-ray technique department was founded in Leningrad Electrotechnical Institute "LETI", although the study in physics of X-ray radiation, design and technology of X-ray devices, as well as the development and application of methods of X-ray examination at that time was conducted in the leading universities of Leningrad – University, Polytechnic and Technological Institutes. They are now also held there, but the department of electronic instruments and devices is still the only one of its kind in Russia.

A considerable role was played by Alexander Stepanovich Popov – the creator of the radio and the first elected director of Leningrad Electrotechnical Institute "LETI". As it is well known, X-ray radiation was discovered by professor of the Würzburg University (Germany) Wilhelm Conrad Roentgen on 8 November 1895. Less than in two months, in January 1896 A S Popov with the assistance of the Mine Officers' Training Course (Kronstadt) worker produces the first in Russia X-ray tube according to the scheme of the Englishman Crookes. And in early February he already demonstrates the experimental production of X-ray images of different objects placed in a cardboard box on the meeting of the Russian physicochemical society in Saint Petersburg University. Alexander Stepanovich was a skilled glass-blower and, for example, made his wife colored perfume bottles. Of course, he could make, as it is said in technology, a vacuum-tight glass container of X-ray tube. It is obvious that research in the field of X-ray technique did not end in his untimely demise, but was successfully continued.

The appointment in 1938 of F N Kharadzha as a head of X-ray technique department, who was a production worker with extensive experience was no less symbolic, and meant at that time a close connection of the educational and research process with the needs of the relevant sector of industry. Satisfaction of the needs of industrial enterprises on the basis of a deep scientific study of the task became a solid basis in the formation and subsequent development of the department and its future.

In this regard, it should be called another name – August Augustovich Potsar, also a graduate of Leningrad Electrotechnical Institute "LETI". He came to the X-ray technique department having extensive practical experience in the measurement laboratory of the "Svetlana" plant. Together with F N Kharadzha he conducted one of the first scientific and technical research on the development of gasotrons. Their result was formulated principles for the design and technology of high-voltage vacuum devices on the basis of the partitioned glass shell. To this day in the nomenclature of products of JSC "Svetlana-X-Ray" is listed the 2 MeV partitioned X-ray tube for industrial inspection. Not by chance as the head of the organized in 1946 at the faculty of electronic technique of Leningrad Electrotechnical Institute "LETI" department of industrial electronics (since 1952, the department of gas discharge devices) was appointed A A Potsar. Subsequently, the fate repeatedly brought together and separated these two departments. Last and final their unification under the name "X-ray and electron beam devices" occurred in 1980. In 1992 the department received its present name – "electronic instruments and devices".

Figure 2. On the sidelines of the Conference.

In confirmation that a close relationship, established from that time, between the department of electronic instruments and devices and domestic electronic industry is not broken, you can name a few names from a very extensive list of executives, and that as the lecturers work with students: Alimov O A (general director of scientific and research enterprise "Electron"), Brytov I A (deputy of the general director of scientific and production enterprise "Burevestnik"), Vetrov N Z (general director of JSC "SED-X-Ray"), Vildgrube G S (general director of scientific and production enterprise "Electron"), Klevtsov V A (deputy of the general director of "Svetlana"), Kulikov N A (general director of JSC "Svetlana-X-Ray"), Komyak A I (general director of scientific and production enterprise "Burevestnik"), Lesish Y K (chief engineer of the Leningrad association of electronic industry "Svetlana"), Stepanov R M (chief engineer of the scientific and production association "Electron"), Tataurshikov S S (chief engineer of the scientific and production association "Electron"), Schukin G A (general director of the Leningrad association of electronic industry "Svetlana").

In 2016, at the 3rd International Conference on X-ray Technique presentations were made by more than 20 specialists of enterprises, actively engaged in research in the field of creation and application of X-ray equipment. Traditionally of special interest were reports of the leading members of the JSC "Svetlana-X-Ray", dedicated to the development of X-ray tubes and betatron cameras of new generation. They are designed for projection radiography, industrial radiography and medical diagnostics and therapy. All the designs were put into mass production and are used in most modern domestic X-ray machines and digital radiography systems. Of particular note is the X-ray tube with a photoelectron multiplication of the anode current, which has no analogues in the world.

For the first time, were so widely presented materials of the younger generation of researchers and students of the department of electronic instruments and devices on the results of research in the field of application of X-ray radiation for medical purposes, including the development of methods for X-ray image acquisition, processing of the obtained images, estimation of radiation dose, etc. Almost every report of graduate students and young scientists of Saint Petersburg State Electrotechnical University "LETI" caused a lively debate, which contributed to an objective assessment of the results.

In summing up the Conference results, it was emphasized that it is very important for the development of domestic production of X-ray tubes, including the preparation of highly qualified specialists. In general, it creates the preconditions for the promotion of Russian industry of X-ray apparatuses production into a leading position in the world.

IOP Conf. Series: Journal of Physics: Conf. Series **808** (2017) 011002 doi:10.1088/1742-6596/808/1/011002

Peer review statement

All papers published in this volume of *Journal of Physics: Conference Series* have been peer reviewed through processes administered by the proceedings Editors. Reviews were conducted by expert referees to the professional and scientific standards expected of a proceedings journal published by IOP Publishing.

Development of digital processing method of microfocus X-ray images

N E Staroverov, E D Kholopova, A Yu Gryaznov and K K Zhamova

Saint Petersburg Electrotechnical University, 197376, Saint Petersburg, Russia

E-mail: nik0205st@mail.ru

Abstract. The article describes the basic methods of X-ray images digital processing. Also in the article is proposed method for background image aligning based on modeling of distorting function and subtracting it from the image. As a result is proposed the improved algorithm for locally adaptive median filtering for which has been carried out the effectiveness experimental verification.

1. Introduction

Micro focus X-ray is a method for producing X-ray images using radiation sources that have focal spot of the size less than 100 μm. The use of these radiation sources makes possible to obtain images with a direct image magnification 5–20 times [1, 2]. Nowadays microfocus X-ray is widely used in medicine, industrial non-destructive testing and evaluation of the crops quality [3]. In some cases, microfocus X-ray images cannot be reliably evaluated due to poor image quality.

The main disadvantages of microfocus X-ray images are irregular background, distorted brightness characteristics and the availability of noises. Due to these disadvantages is required to carry out microfocus X-ray images digital processing. To reduce these shortcomings in the article are proposed processing technique of microfocus X-ray images. The proposed method of image processing consists of two stages: image background correction and noise suppression based on locally adaptive median filtering.

2. Image background correction

During the X-ray examination by means of the scheme with the image zooming, appears irregular image background. This fact complicates the visual analysis and subsequent digital image processing. To eliminate the uneven image background is introduced the distorting function. It is assumed that the receiver records the image, which can be represented as the sum of image and distorting function that creates an irregular background. In the case of microfocus X-ray distorting function usually can be described by a function of the form:

$$g(x,y) = C\left[\left(x - \frac{x_{\max}}{2}\right)^2 + \left(y - \frac{y_{\max}}{2}\right)^2\right], \tag{1}$$

where C – is a constant and its value depends on the microfocus X-ray machine mode and the operating conditions. This type of distorting function occurs because of shooting features of image zooming, since the intensity of X-ray is reduced in proportion to the square of the distance. The following method can be used to determine the constant C. By the Sobel method in the image are

IOP Conf. Series: Journal of Physics: Conf. Series **808** (2017) 012001 doi:10.1088/1742-6596/808/1/012001

allocated boundaries of the object, then is performed the operation of holes morphological fill in. After that are defined the image edges profiles, the profiles are based only on the parts of the image edges that do not belong to the objects. For each profile are determined minimum and maximum, and then is calculated the difference between them. The constant C is determined as the average difference between the profiles minimum and maximum. The image $f(x, y)$ is obtained by subtracting $g(x, y)$ from the registered image:

$$f(x, y) = h(x, y) - g(x, y). \qquad (2)$$

Figure 1 shows the X-ray image of BGA chips, as well as the image background surface. Figure 2 shows the result of subtracting the distorting function $g(x, y)$ from the X-ray image.

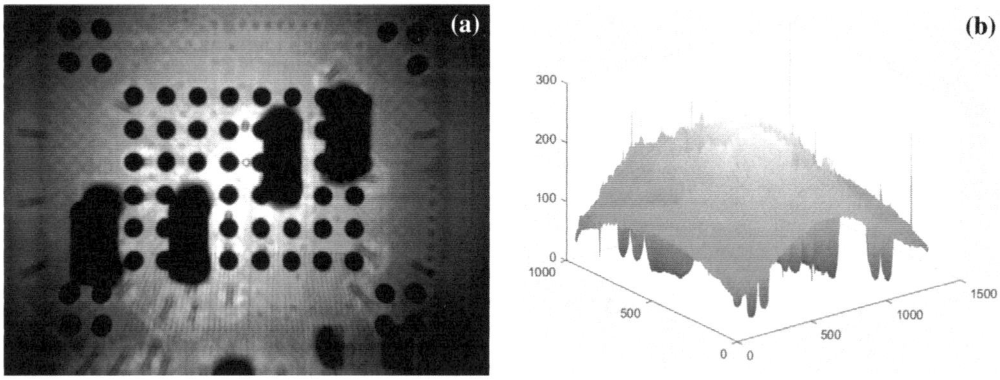

Figure 1. X-ray image of BGA chip (**a**) and the surface of the image background (**b**).

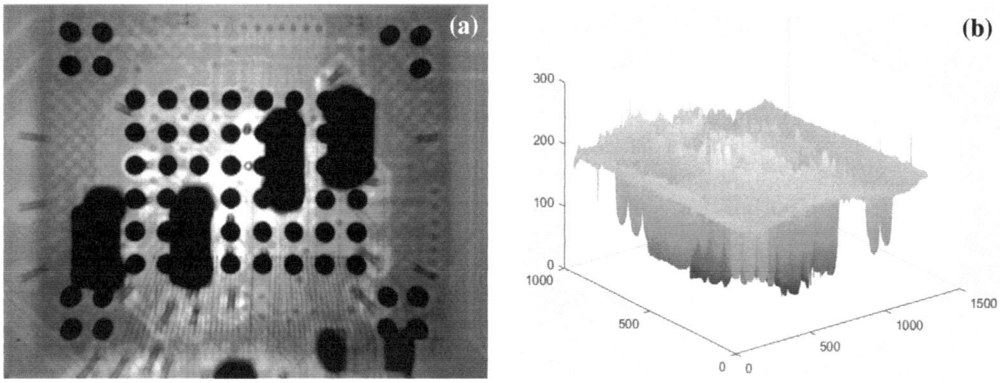

Figure 2. X-ray image of BGA chip after background correction (**a**) and the surface of the image after the background correction (**b**).

As can be seen in figure 2, the image surface is leveled; uneven illumination of the background is removed. After correction of image background it is much more convenient to carry out the automatic analysis, in addition it is easier to carry out visual analysis. For example, for the represented image the chip pads can be identified by simply applying a threshold conversion. In case of the non-uniform background, this conversion is inefficient.

3. Median filtering

Today to reduce the noise on X-rays image are often used averaging filter or filters based on statistical processing (median filter). However, these methods have significant disadvantages: weak suppression of impulse noise and components distortion in the image. As an example of the median filtering is shown in figure 3(b). We can see that this filter provides insufficient noises suppression, as well as it degrades sharpness. The modified locally adaptive median filter, which is proposed in [4] has no such disadvantages.

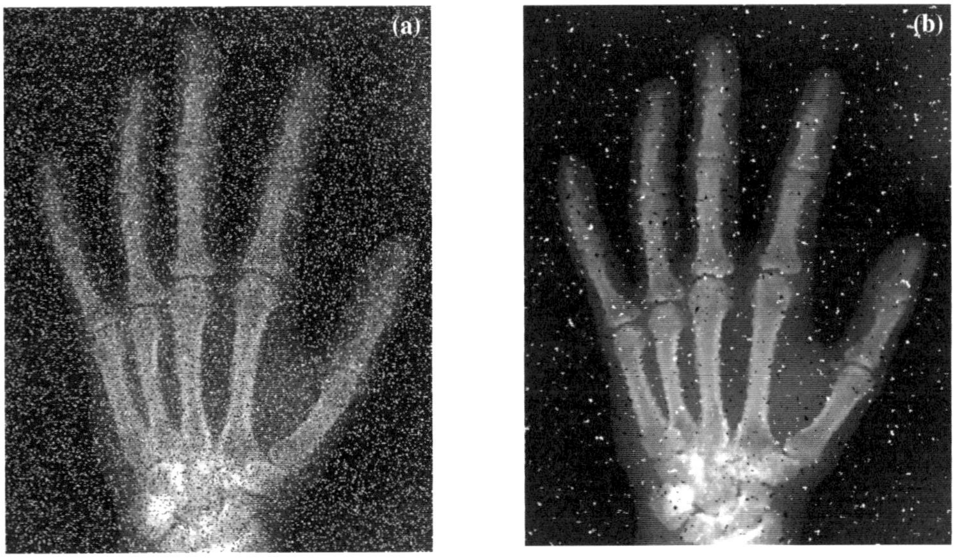

Figure 3. The original image (**a**) and the image after filtering by median filter size of 7×7 pixels (**b**).

Locally adaptive median filter operates on the following algorithm:

Part 1:

$A_1 = Z_{med} - Z_{min}$; $A_2 = Z_{med} - Z_{max}$;

if $A_1 > 0$ and $A_2 < 0$, to part 2; otherwise increase S;

if $S \leq S_{max}$, repeat part 1; otherwise the result is Z_{med}.

Part 2:

$B_1 = Z_p - Z_{min+1}$; $B_2 = Z_p - Z_{max-1}$;

if $B_1 > 0$ and $B_2 < 0$ the result is Z_p; otherwise the result is Z_{med},

where Z_{min} and Z_{max} are the minimum and maximum brightness values in the vicinity of S; Z_{med} – median brightness in the vicinity of S; Z_p – brightness value of the pixel being processed; S_{max} – maximum size of the area. The algorithm works as follows. In the first part is carried out the check whether the median is of local minimum or maximum. If it is, the size of the area is increased, in the opposite case is carried out transition to part 2. In part 2 is checked whether the value of pixel brightness is of maximum or minimum in the area. If "yes" – the brightness of a pixel is changed to the median, if not, it remains unchanged.

The considered filter suppresses noise and retains much more image details than ordinary median filter. However, when filtering the micro focus X-ray images it has been found that in some cases, the filter removes noise less effectively, for example, in case when in the vicinity is located more than one noise pixel. The example of locally adaptive median filter is shown in figure 4(a). This filter does not substantially degrade image sharpness, though it does not also fully suppress the noise. This can occur if in the image receiver ceases to work closely located spaced pixels or if on the luminescent screen system of a digital camera fall fine metal particles.

To improve the discussed filter was proposed to replace in part 2 minimum and maximum brightness value in the vicinity of S for the brightness value after reaching a minimum and to the maximum (second minimum and a second maximum). Then part 2 will look like:

$B_1 = Z_p - Z_{min+1}$; $B_2 = Z_p - Z_{max-1}$;

if $B_1 > 0$ and $B_2 < 0$ the result is Z_p; otherwise the result is Z_{med}.

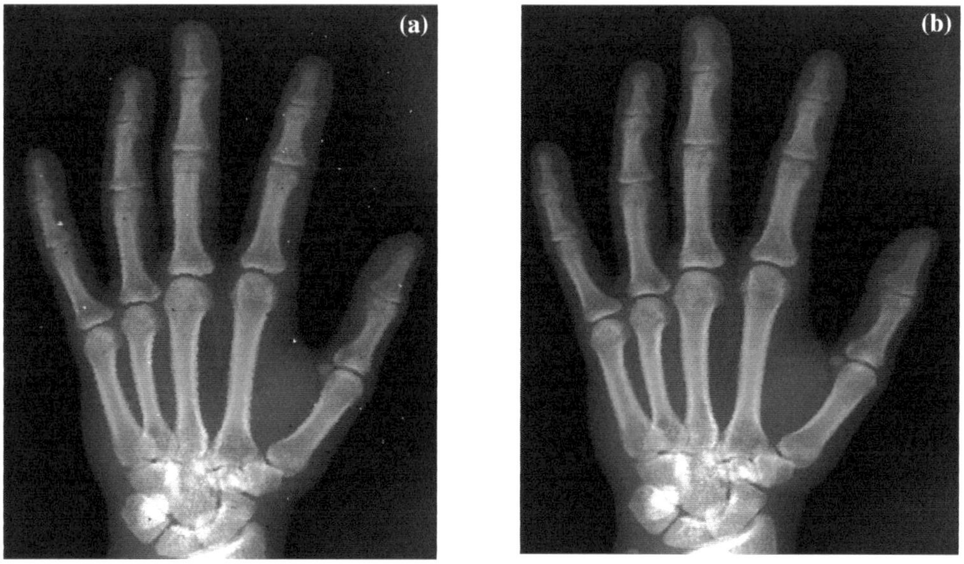

Figure 4. Image after filtering with the locally adaptive median filter (**a**) and the image after filtering by the improved locally adaptive median filter (**b**).

4. Conclusions

The studies have shown that modified filter effectively suppresses noise and slightly reduces the sharpness of the X-ray image. The proposed method allows correcting the background of micro focus image without using background image and without operator. In addition, it also allows removing image noise without compromising image sharpness and without distorting of fine detail. The analysis of images processed by the proposed method shows that processed images are more informative than processed by conventional techniques.

References

[1] Potrakhov N N 2007 *Vestnik novykh meditsinskikh tekhnologiy* **3** 167–9

[2] Potrakhov N N, Gryaznov A Yu, Zhamova K K, Bessonov V B, Obodovski A V, Staroverov N E and Kholopova E D 2015 *Biotechnosphere* **5** 46–54

[3] Gryaznov A Yu 2010 *Methods and systems of microfocus phasecontrast X-ray medical diagnosis* (Saint Petersurg: Saint Petersburg Electrotechnical University)

[4] Gonzalez R C and Woods R E 2008 *Digital Image Processing* (New Jersey: Prentice Hall)

IOP Conf. Series: Journal of Physics: Conf. Series **808** (2017) 012002 doi:10.1088/1742-6596/808/1/012002

TXRF spectrometry at ion beam excitation

V Egorov[1], E Egorov[1] and M Afanas'ef[2]

[1] Institute of Microelectronics Technology and High Purity Materials, Russian Academy of Sciences (IMT RAS), 142432, Chernogolovka, Moscow Distr., Russia
[2] Institute of Radio-engineering and Electronics, Russian Academy of Sciences (IRE RAS), 125009, Moscow, Russia

E-mail: egorov-iptm@mail.ru

Abstract. The work presents short discussion of TXRF and PIXE methods peculiarities. Taking into account of these peculiarities we elaborate the experimental scheme for TXRF measurements at ion beam excitation of characteristical fluorescence. The scheme is built on base of the planar X-ray waveguide-resonator with specific design. Features of the new experimental method and possibilities of Sokol-3 ion beam analytical complex were used for the method application in real measurements.

1. Introduction

X-ray fluorescence analysis in conditions of the total external reflection of the exciting radiation flux on the studied material surface (TXRF) is today the most effective method for quantitative elemental diagnostical material objects on base of X-ray fluorescence application [1]. The method realization in the total external reflection conditions provides excitation of a surface layer with thickness 3–5 nm. So, X-ray fluorescence yield excited in these conditions characterizes the element composition of thin surface layer. Moreover, the background deposit into TXRF spectrum has very small value, and the spectrum dispenses with the need for the matrix correction [2].

TXRF application allows decreasing of pollution detection limits on 1.5–2 orders in comparison with XRF conventional measurements. At the same time, TXRF analysis as well as conventional X-ray fluorescence spectrometry at excitation of the fluorescence yield by X-ray fluxes, gamma radiation and electron beams has some difficulties at the light elements diagnostics in materials. It is illustrated by figure 1 presenting cross-sections of K_α and L_α characteristical fluorescence excitation by MoK_α radiation flux for different elements. YK_α and AlK_α cross-section comparison shows that the fluorescence yield of these lines is different on four orders.

Fluorescence difficulties of light element diagnostics in materials can be compensated by ion beam excitation of X-ray fluorescence yield (by PIXE method) [3]. The mechanism of fluorescence excitation by ion beams differs from ones featured to other excitation methods. PIXE bears similarly to the fluorescence yield produced by electron excitation but it is characterized by small value of the bremsstrahlung photon intensity yield. In the result, PIXE spectrometry demonstrates low levels of the background in comparison with the electron beam excitation of the fluorescence yield. But the main preference of the PIXE method is high cross-section of X-ray fluorescence excitation for light elements. Experiments showed that proton and helium ion beams with energy near 1 MeV excite the light element fluorescence more effective in comparison with the fluorescence of heavy elements [4]. Figure 2 allows to compare the cross-section of X-ray fluorescence excitation for AlK_α and YK_α lines in condition of proton beam (E_0 = 1 MeV) application. It is apparent that AlK_α line excitation is effective on three orders approximately.

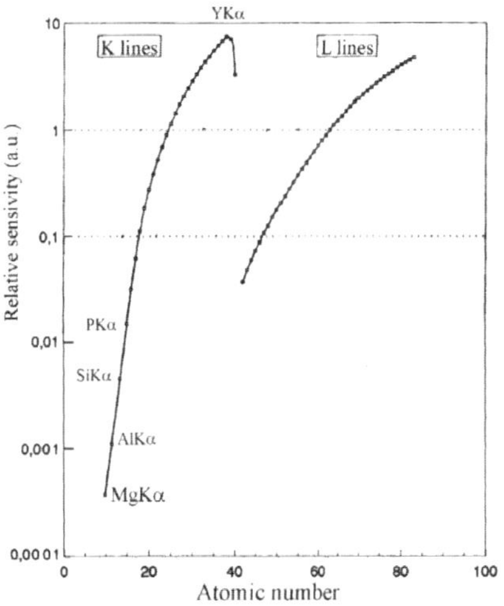

Figure 1. Cross-sections excitation for K_α and K_β different elements lines yields in conditions of MoK_α radiation flux using as the exciting agent [1].

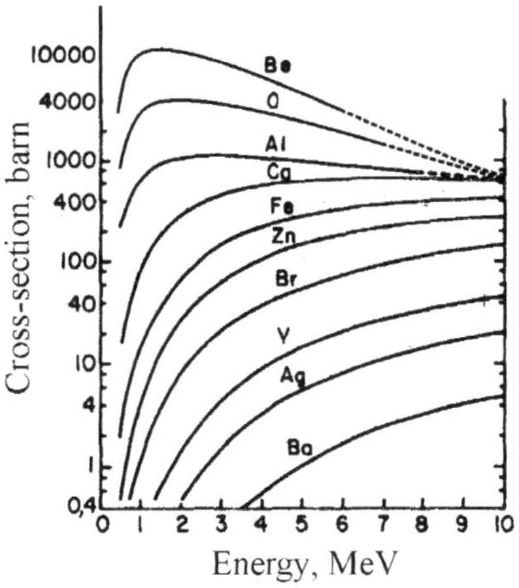

Figure 2. Energy dependences of the cross-section excitation for K_α different elements lines yield in conditions of H^+ ion beam using of the exciting agent [4].

2. PIXE method in TXRF condition

TXRF spectrometry registries of X-ray fluorescence yield from thin film surface layer. Unlike TXRF method PIXE spectrometry is characterized by registration of the total fluorescence yield excited in the layer, which thickness is defined by the ion beam penetration into volume of the studied material. Moreover, the real yield of the characteristic fluorescence will be defined by absorption factor featured for the material. So, every lines will characterize the layer with thickness defined by own absorption factor. TXRF method is free from the matrix influence. PIXE measurements must take into account this factor. Suppression of the matrix factor influence is not simple task. At the same time, removal of the fluorescence yield dependence from layer absorption thickness is not difficult problem. This task can be solved by application of the specific X-ray optical scheme for X-ray fluorescence yield registration executed on base of the planar X-ray waveguide-resonator (PXWR) with hole in its central position (figure 3).

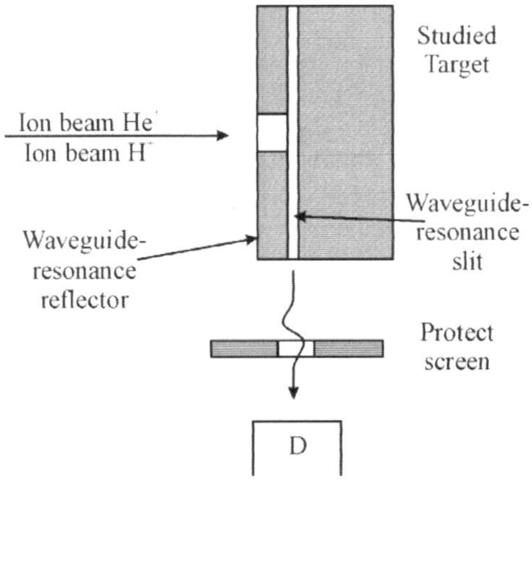

Figure 3. Scheme of X-ray fluorescence yield corresponding to thin surface layer allowed using ion beam excitation in conditions of the planar X-ray waveguide-resonator application. PXWR slit width 0.15 μm.

This scheme allows forming the waveguide-resonance structure by surface of the studied material and the polished surface of Be reflector carried Ti strips on its edges with thickness 0.15 μm. In the result, we have the waveguide-resonance slit clearance for X-ray fluorescence lines of light elements. Proton ion beam incidents on the target through hole in the Be reflector and excites the target characteristic fluorescence. At the same time, X-ray optical scheme provides registration of X-ray fluorescence yield for thin surface layer of the target with thickness 3–5 nm. This registration scheme is hybrid of PIXE and TXRF investigation methods but it does not deliver from the matrix effects influence.

3. TXRF in the PIXE conditions

TXRF spectra collected in conditions of ion beam fluorescence excitation is the best support for RBS spectra approximation for material involving light elements. Figure 4 demonstrates RBS spectrum of H^+ ion beam (E_0 = 1.18 MeV) for multielement target containing of the light element set. Spectrum shows some vagueness in Al, Si and P atomic concentration owing to not enough of its energy resolution. In this case the TXRF information obtained at ion beam excitation is very useful. Figure 5 shows X-ray fluorescence yield spectrum of the multielement target collected on condition of excitation by proton beam with energy E_0 = 1.18 MeV. The spectrum demonstrates SiK_α, AlK_α and NaK_α intensity lines, envelope of PK_α and ZrL_α composite line and small peak of OK_α. This spectrum approximation allows removing some vagueness at RBS spectrum interpretation and makes it possible to determinate element composition of the multielement target. It is $Na_3Zr_{1.3}Si_{1.9}Al_{0.1}P_{1.0}O_{12}C_2$. It is important to notice that the PIXE spectrum collected in the conventional conditions showed perceptible differences in comparison with spectrum presented in figure 5. It is suggested that the comparison of TXRF spectra collected in conditions of ion beam and X-ray excitations permits to elaborate the objective method allowed to evaluate the matrix effect influence on TXRF spectra ion beam excitation.

4. Conclusion

The work presents short discussion of TXRF and PIXE method and suggest new procedure of TXRF and PIXE hybrid investigation. It is shown that the new method can be used for investigation of multielement targets containing light element set.

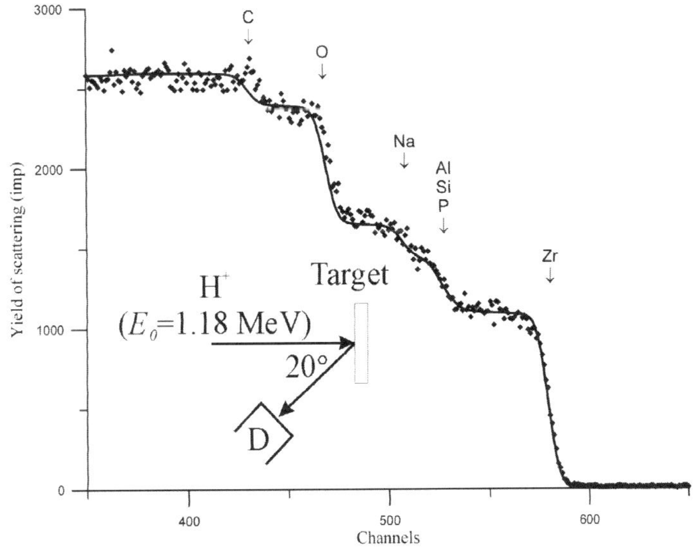

Figure 4. Experimental and theoretical spectra RBS H^+ ion beam (E_0 = 1.18 MeV) for multielements target collected for θ = 160° scattering angle. Arrows show energies corresponding to scattering on nuclears of surface atoms. Energy step 1.9 keV/channel.

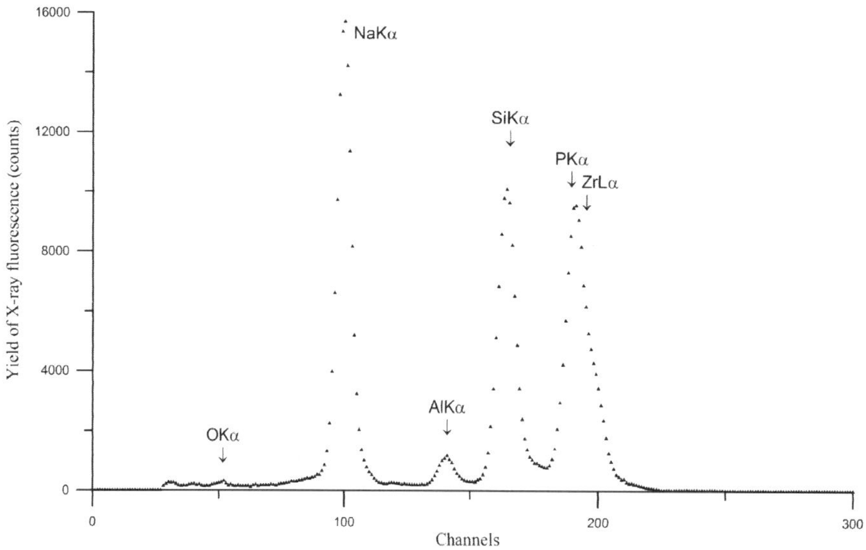

Figure 5. TXRF spectrum collected for multielement target in conditions of excitation by H^+ ion beam (E_0 = 1.18 MeV) at PXWR with specific design application. PXWR slit width was 0.15 μm. SSD registrator was equipped by Be window 8 μm. Energy step 10.2 eV/channel.

Acknowledgements
This work was partially supported by Russian Foundation for Basic Research (projects #15-08-02618, 16-07-00665).

References

[1] Klockenkamper R 2014 *Total reflection X-ray fluorescence analysis* (New York: Wiley)
[2] Van Grieken R and Markowicz A A 2002 *Handbook of X-ray spectrometry* (New York: Dekker)
[3] Johanson S A E, Campbell J L and Malqvist K G 1995 *Particle induced X-ray emission spectrometry (PIXE)* (New York: Wiley)
[4] Cahill T A 1980 *J. Annu. Rev. Nucl. Part. Sci.* **30** 211

IOP Conf. Series: Journal of Physics: Conf. Series **808** (2017) 012003 doi:10.1088/1742-6596/808/1/012003

Application of operational radiographic inspection method for flaw detection of blade straightener from polymeric composite materials

A N Anoshkin[1], V M Osokin[1], A A Tretyakov[1], N N Potrakhov[2] and V B Bessonov[2]

[1] Perm national research polytechnic university, 614000, Perm, Russia
[2] Saint Petersburg Electrotechnical University "LETI", 197376, St. Petersburg, Russia

E-mail: osokin-kt@pstu.ru

Abstract. In the article on the example of the straightener blade made of polymer composite materials, discusses the advantages of using the method of microfocus X-ray for nondestructive testing of aviation products. Described basic types of defects characteristics occurring in a similar type parts both during their manufacture and during their operation, namely, interlayer delamination, pores and wrinkles. Peculiarities of microfocus X-ray are shown, which is the use of radiation sources with a focal spot size of less than 100 μm. These features make it possible to increase the details and therefore, to minimize the size of detected defects in transmission. On the basis of experimental studies were defined radiographic signs of major types of defects, typical for products made of polymeric composite materials. Calculated time costs of personnel required for high-resolution X-ray recording and evaluation of test results.

1. Introduction

Currently, the aircraft engine-engineering global trend is to replace products made of metal alloys to their counterparts made of polymeric composite materials (PCM). This allows achieving significant weight reduction, fuel efficiency, and in some cases, manufacturability improvement and reduction of the manufacturing costs. Operating experience of parts and components of PCM shows that fracture typically occurs as a result of delamination of monolayers in places of defects localization, although the material still has a sufficient margin of safety in the direction of reinforcement. Thus, the transition to the production of highly loaded propulsion elements from composite materials requires careful solution of problems of internal flaws reliable detection. The main types of defects occurring both in manufacturing process of parts and in their operation, are interlayer delamination, pores and wrinkles [1]. Delamination in construction may occur at comparatively low loads well before the exhaustion of its load-bearing capacity. In some cases, they have a very small opening (less than 0,1 mm), which affects the accuracy and reliability of their identification. In this paper identification of these defects is investigated on the example of the straightener blade (SB) (figure 1), made of carbon fiber by autoclave forming in the mold [2, 3].

2. Methods and equipment of microfocus X-ray

Currently, one of the most informative and accurate non-destructive testing methods to identify typical defects is a microfocus X-ray method, used in the present work. Microfocus X-ray is a method of

IOP Conf. Series: Journal of Physics: Conf. Series **808** (2017) 012003 doi:10.1088/1742-6596/808/1/012003

obtaining X-ray images of various objects using the X-ray source with focal spot size less than 0,1 mm [4]. Unlike traditional methods of X-ray examinations, microfocus X-ray has a number of important features. The first of these is the ability to obtain sharp magnified X-ray images of various objects.

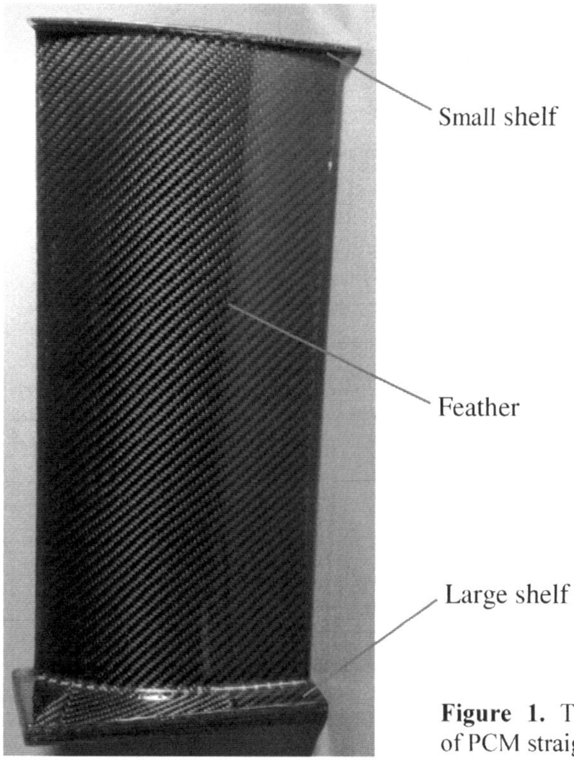

Figure 1. The appearance of PCM straightener blade.

Obtaining increased X-ray imaging is one of the important conditions for nondestructive testing of PCM products. This is due to close values of density of material, reinforcing elements and binder, resulting in low contrast images thereby reducing the probability of detection and identification of possible defects using the traditional radiography. At the same time focal spot size, and the distance between the radiation source and the subject substantially affect the value of the geometric component, i.e. the blur of the image. The second, equally important, feature of the microfocus X-ray is the ability to obtain informative X-ray images the same research objects with less radiation load compared to classical X-ray [5]. In this regard, to solve the problem of highly informative operational control of PCM SB was used method of microfocus X-ray imaging.

Equipment for the microfocus X-ray used in this study, includes portable microfocus X-ray diagnostic complex RAP-150 (figure 2(a)) developer and manufacturer of which is a CJSC «Elektronnaya Tekhnica-Meditsina». Microfocus X-ray machine is made in monoblock design, the anode voltage of the device does not exceed 130 kV, anode current is less than 200 μA, focal spot size ranges from 20 to 100 μm. The storage plate with photo-stimulated phosphor were used as a digital X-ray image detectors. Reading of the plates was performed using the CR 35 NDT scanning system (figure 2(b)), which has a resolution of 15 line pairs per mm.

3. Results and discussion

During the implementation of this research was carried out testing of the microfocus X-ray inspection method on the example of PCM SB. Figures 3 and 4 show examples of magnified X-ray images of detected defects in the form of pores in the feather part of the SB, as well as in transition from feather

IOP Conf. Series: Journal of Physics: Conf. Series **808** (2017) 012003 doi:10.1088/1742-6596/808/1/012003

part to shelf. A result of researches were determined the minimum detectable defects, which do not exceed 0.5 mm.

Figure 2. Microfocus X-ray machine (**a**) and CR 35 NDT scanning system (**b**).

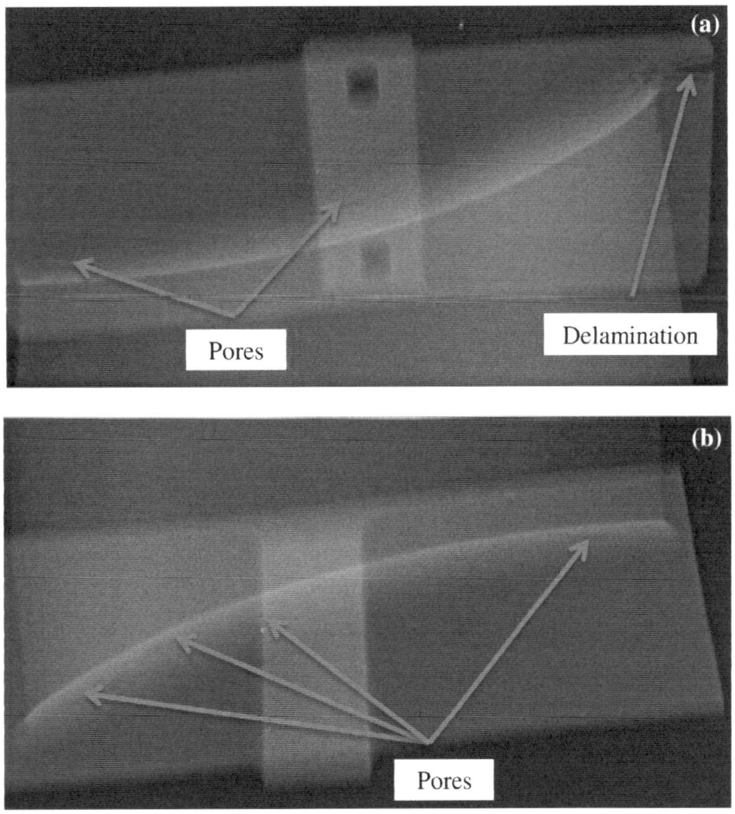

Figure 3. Examples of X-ray images of the PCM SB with the identified defects in large (**a**) and small shelves (**b**).

At the modern stage of medical engineering, the main problem related to X-ray tubes is obtaining high-intensity X-radiation with ensuring small size of focal spots. This is especially important for X-ray tubes which are parts of computerized tomographic scanners, angiographers, mammographic scanners and digital scanning X-ray devices.

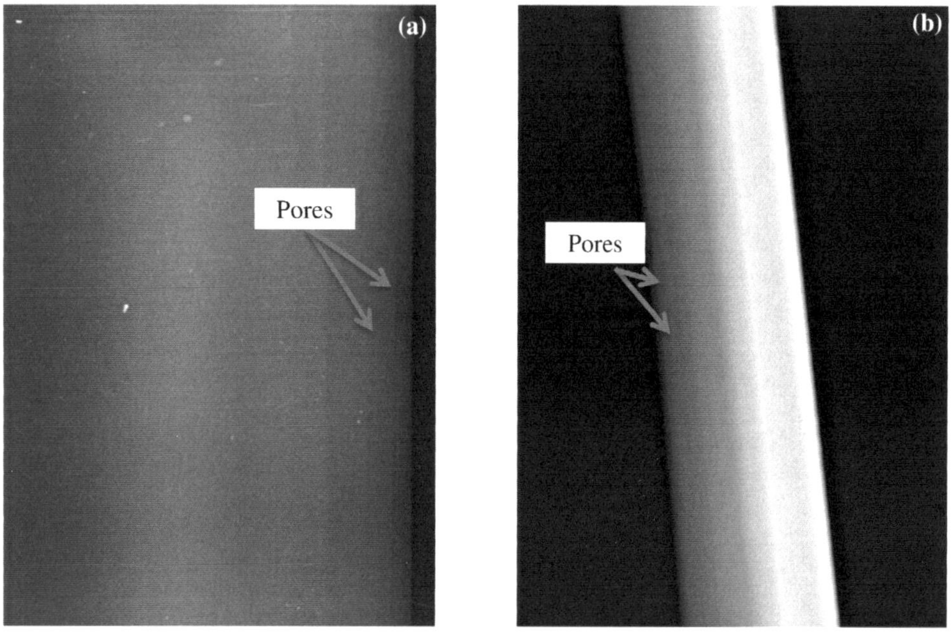

Figure 4. Examples of X-ray images of the PCM SB with the identified defects in feather part.

4. Conclusions

The results of using technologies for efficient X-ray control of the PCM SB leads to the conclusion that, technology enables detection of main types of defects located both in the shelves of the blade straightener and its feather part, as well as in bending areas of layers, i.e. the transition from the feather to shelves. At the same time the introduction of X-ray inspection method for flaw detection of PCM SB allows to reliably detect defects, located in inaccessible areas for other classic non-destructive testing methods. One of the main such place in the blade design is bending of layers, i.e. the transition from the feather to small and large shelves. Obtaining time of X-ray image does not exceed 1 min, and the total time of the analysis taking into account the scanning of storage plates and visual assessment by operator is 5–7 min.

Acknowledgment

This work was financially supported by the state in the face of Russian Ministry of Education in the framework of the federal target program «Research and development in priority areas of Russian scientific and technological complex for 2014-2020» related to «Scientific substantiation of design and technological solutions for creation of highly-loaded components of advanced aircraft engines, subjected to intense aerodynamic effects, made of polymeric composite materials on the example of the blade straightener». Unique identifier of applied scientific research RFMEFI57414X0080.

References

[1] Potrakhov N N, Zhamova K K, Bessonov V B, Gryaznov A Y and Obogovsky A V 2015 *PNRPU Aerospace Engineering Bulletin* **43** 97–115

[2] Gagauz F M 2012 *The design and production of aircraft constructions. Collection of scientific papers KhAI* **4** 15–20
[3] Anoshkin A N, ZuikoV Yu, Shipunov G S and Tretyakov A A 2014 *PNRPU Mechanics Bulletin* **4** 5–44
[4] Rumyantsev S V 1974 *X-ray flaw detection* (Moscow: Atomizdat)
[5] Potrakhov N N and Gryaznov A Yu 2012 *Microfocus X-ray imaging in medical diagnostics* (Saint Petersburg: Saint Petersburg Electrotechnical University)

IOP Conf. Series: Journal of Physics: Conf. Series **808** (2017) 012004 doi:10.1088/1742-6596/808/1/012004

Cathodes for medical purpose X-ray tubes

M L Taubin[1,2], D A Chesnokov[1] and A A Pavlov[1]

[1] Federal State Unitarian Enterprise Scientific Research Institute the Scientific Industrial Association "Luch", 142100, Moscow region, Russia
[2] National Research Nuclear University "MEPhI", 115409, Moscow region, Russia

E-mail: taubin@sialuch.ru

Abstract. Results of own works of the authors and works of the Russian, Japanese, Korean and American experts in creation nanocomposites of refractory metals with increased strength characteristics for cathodes are analyzed in this report. It is shown, that though monocrystals have shown the better characteristics in comparison with traditional polycrystals as cathodes, more radical way of increase of operational characteristics of the X-ray tubes is application refractory nanostructural materials. Some results of investigations in which the work function electrons of nanocrystal tungsten is lower by 0.8 eV than that of traditional tungsten are given. This effect is able to increase the intensity of X-radiation considerably (by a factor of more than 5) at the same cathode temperatures or decrease the temperature of a cathode by 400 °C at the same intensity of X-radiation.

At the modern stage of medical engineering, the main problem related to X-ray tubes is obtaining high-intensity X-radiation with ensuring small size of focal spots. This is especially important for X-ray tubes which are parts of computerized tomographic scanners, angiographers, mammographic scanners and digital scanning X-ray devices.

At the same time, the problems related with cathodes make the key issue regarding obtaining the desired X-ray optical characteristics of the X-ray tubes. In this aspect, focusing of electron beam is the crucial for obtaining focal spots of the desired sizes. Presently, the most promising solution for obtaining small-sized focal spots is to utilize flat emitting surface cathodes (figure 1).

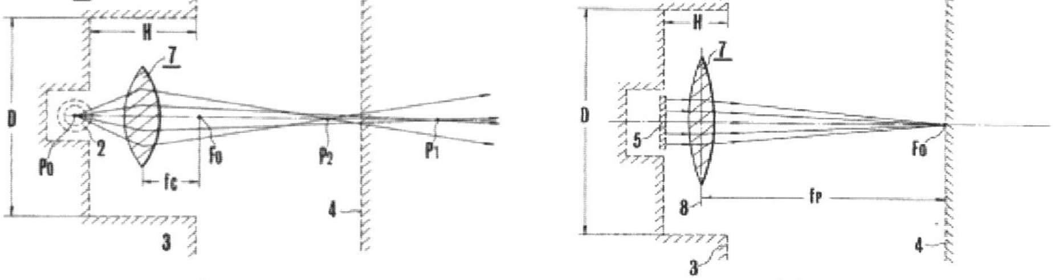

Figure 1. Schematic of electron beam focusing from the cylindrical surface of a spiral (**a**) and a flat surface (**b**) [1]: 1 – cathode; 2 – emitter; 3 – focusing head; 4 – object.

Flat-shaped emitter (figure 2) is more efficient than a spiral-shaped one because of having more even electric field distribution, and furthermore – a greater surface area which is also important for obtaining a more intense X-radiation [2, 3].

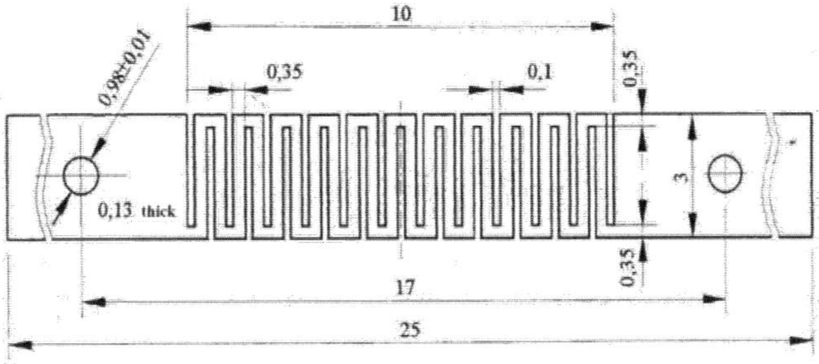

Figure 2. Drawing of a flat emitter (variant).

In addition to the promising design features, intensity of the electron beam is also important. This parameter is mainly determined by the work function of the emitting material. To increase the emission current intensity, the work function can only be lowered by using new materials.

The results of longstanding studies of materials regarding their use as emitters of thermionic transducers [3] showed that monocrystalline tungsten may also become a promising material for X-ray tube emitters. The results of these studies were indicative of higher emission characteristics, lesser and more uniform tungsten vaporization from the surface as compared with the polycrystalline structures as well as higher dimensional stability at high operational temperatures due to the lack of recrystallization (which is a significant shortcoming of polycrystalline emitters) up to the melting point. For the foregoing reasons, development of the flat emitter cathode made of monocrystalline tungsten seems to be a promising trend.

Utilization of monocrystalline tungsten as the emitter material makes it possible to:

1) reduce the emitter operational temperature by 70 °C since crystallographic (111) plane of monocrystalline tungsten has work function of (4.4 ± 0.05) eV, which is 0.15 eV lower than that of polycrystalline tungsten (4.55 ± 0.05) eV;

2) facilitate electron beam focusing due to more uniform emission of the monocrystalline structure.

Experimental studies of emissive properties (figure 3) confirmed that monocrystalline tungsten with emitting plane crystallographic orientation (111) has higher emission characteristics as compared with polycrystalline tungsten.

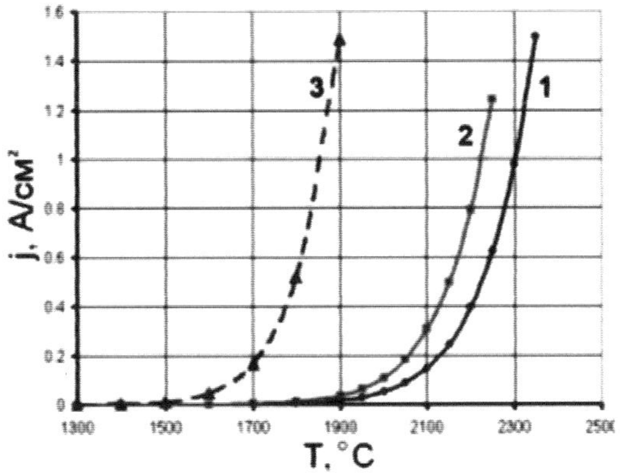

Figure 3. Emission current density versus temperature dependence for: 1 – polycrystalline tungsten; 2 – monocrystalline tungsten (emission from crystallographic plane (111)); 3 – nanocrystalline tungsten (calculations using the results of article [4]).

The second investigated parameter was vulnerability of the monocrystalline structure under exposure to the electron beam. It was shown that X-ray tube dosage rate reduction (vulnerability parameter) for the monocrystalline structure (figure 4) is somewhat different from the typical behavior of the polycrystalline material, and the monocrystalline X-ray tube cathode is more stable throughout the service life. Thus, after dosage rate reduction for 10 % during the first 1500 runs the system becomes stabilized, resulting in a longer service life of the X-ray tube.

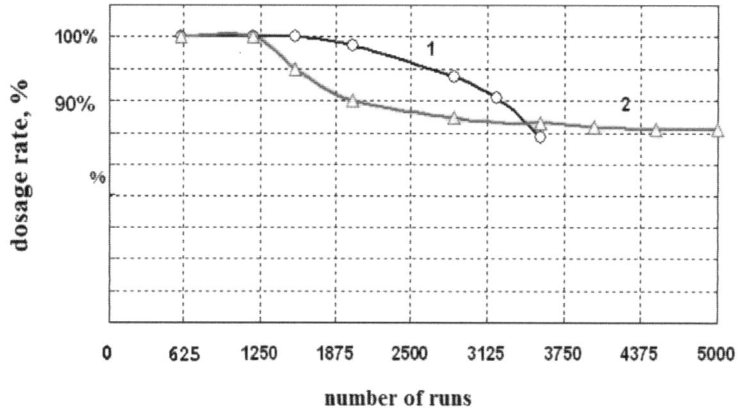

Figure 4. Dosage rate versus number of runs dependence: 1 – rhenium-doped polycrystalline tungsten; 2 – monocrystalline tungsten.

Monocrystalline materials as cathode emitters also performed better in terms of evaporation capability (figure 5), which is of especial significance with account of high operational temperature of the emitter (2400 °C).

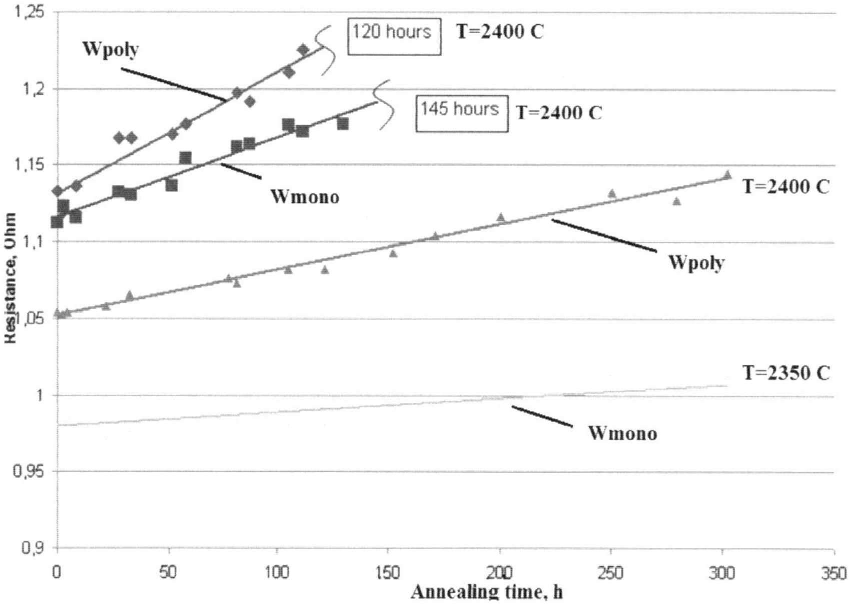

Figure 5. Comparison of evaporation capacity of tungsten with monocrystalline and polycrystalline structures.

Despite the obvious advantages of monocrystalline materials in this context, an even more radical way to enhance the operational characteristics of X-ray tubes is utilization of nanomaterials. According to certain research findings [4], the work function of nanocrystalline tungsten was 0.8 eV lower than that of coarse-grain tungsten. This can be explained by formation of current tube with reduced work function in 10 nm zone close to grain boundaries in the process of nanostructure forming. This may result in significant boost of X-radiation intensity at the same temperature or enable to reduce the cathode temperature for 400 °C keeping the X-radiation intensity at the same level. According to the estimations, at the temperature where the effect improving the emission characteristics of this material still exists ($T \leq 1500$ °C), obtaining the emission current equal to 10 mA/cm^2 is possible.

Recently, a number of articles were published [5–8], providing description of the results of field-emission X-ray tubes development works using carbon nanotubes (CNT) as cathode materials. Thus, the researchers from the South Carolina State University together with the specialists of Xintek Inc. (USA) presented the materials on development of a scanning X-ray source with CNT emitter [6].

Concurrently with enhancing the characteristics of CNT cathode, utilization of nanosized dopants may probably improve mechanical properties of conventionally used tungsten. We managed to improve the mechanical characteristics (figure 6) in the process of high-temperature annealing of samples made of 100 µm thick VA-grade tungsten foils (tungsten with silicon-aluminum and alkali dopants).

Examination of these materials structure after annealing (figure 7) enabled to detect formation of nanosized (≤ 200 nm) inclusions at the grain boundaries of the preannealed tungsten, resulting in a sharp (more than threefold) decline of deformation of the material at a temperature of 2200 °C.

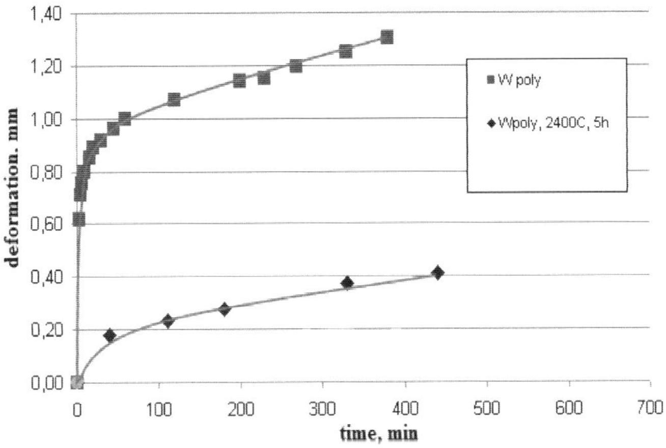

Figure 6. Strain versus time dependence for tungsten at 2200 °C: 1 – polycrystalline tungsten without thermal treatment; 2 – polycrystalline tungsten pre-annealed at 2500 °C for 5 h.

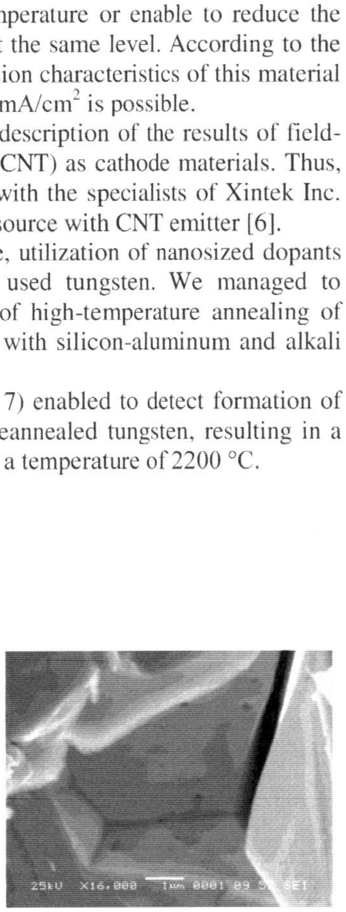

Figure 7. Microstructure of tungsten after thermal treatment at 2400 °C for 5 h (nano-sized pores).

Generally, qualitative improvement of X-ray tube operational characteristics by using nanomaterials and monocrystalline materials may be achieved by simultaneous increase of the electron beam density (flat cathode made of nanostructured tungsten or tungsten nanocomposite) and increase of the X-ray tube output capacity (anode made of monocrystalline tungsten, molybdenum-niobium or tungsten-rhenium nanocomposite) thanks to a higher intensity of interaction between the high-density electron beam and a greater number of atoms of the target anode material. The results presented provide substantiation for development of a new generation X-radiation sources with enhanced diagnostic capabilities.

Conclusions

The authors carried out some experiments to determine the of tungsten deformation. The results of these experiments showed that high-temperature sample annealing resulted in a sharp increase (by a factor of more than 3) in material deformation at a temperature of 2200 °C. The authors think that this effect results from forming nanoscale bubbles filled with potassium at the grain boundaries.

References

[1] Alekseev S V, Taubin M L and Yaskolko A A 2014 *Nanocomposites in X-ray technique* (Moscow: Technosphera)
[2] Kim M 1966 *Metallic materials for vacuum tubes*
[3] Platonov V F, Taubin M L and Hans K 2006 Patent RU 2314592. Directly Heated Cathode and Manufacturing Process
[4] Klinkov A E, Nikolaev Yu V and Vybyvanetz V I 1991 *Proceedings of the Second International Conference – "Nuclear Power Engineering in Space. Physics of Thermionic Energy Converters" (Sukhumi)* p 37
[5] Mulyukov R R and Yumaguzkhin Yu M 2004 *Doklady Akademii Nauk* **399** 760–1
[6] Senda S, Sakai Y, Mizuta Y, Kita S and Okuyama F 2004 Appl. Phys. Lett. **85** 5679
[7] Zhang J, Yang G, Cheng Y, Gao B, Qiu Q, Lee Y Z, Lu J P and Zhou O 2005 Appl. Phys. Lett. **86** 184104
[8] Taubin M L, Platonov V F and Yaskolko A A 2009 *Meditsinskaya Tekhnika* **1** 44–7

IOP Conf. Series: Journal of Physics: Conf. Series **808** (2017) 012005 doi:10.1088/1742-6596/808/1/012005

Thickness measurements of coatings formed from metal plasma of a vacuum arc discharge using X-ray radiation

A A Lisenkov[1], S A Martsinukov[2], V V Chernigovskiy[2], D K Kostrin[2] and E A Smirnov[2]

[1] Institute of Problems of Mechanical Engineering, Russian Academy of Sciences, 199178, Saint Petersburg, Russia
[2] Department of electronic instruments and devices, Saint Petersburg Electrotechnical University "LETI", 197376, Saint Petersburg, Russia

E-mail: lisran@yandex.ru

Abstract. Using vacuum arc plasma sources allows obtaining coatings with broad spectrum of functional properties. The main problem is to control the thickness of a formed coating. This paper shows the possibility of using X-ray radiation to determine the coating thickness.

Currently a wide range of various by composition coatings deposited from metal plasma of a vacuum arc discharge is developed and applied [1, 2]. Universal properties of the formed coatings allow their use in various branches of engineering to improve wear resistance, reduce coefficient of friction and provide protection from corrosion [3, 4]. To meet market requirements the coating must provide a sufficient hardness, low friction coefficient and high thermal stability. The solution to this problem is achieved either by the development of new coating formulations with enhanced performance properties, or by developing new methods of coating deposition to improve the coating quality and increase the speed of its formation. To achieve this goal vacuum arc plasma sources are very widely used [5]. Modifying parameters of a generated flux of metal plasma, the pressure of the reaction gas and the magnitude of the accompanying magnetic field, it is possible to control the properties of condensate and to obtain the necessary coating quality.

In vacuum arc plasma sources with coaxial design of the electrodes the cathode is made in a form of a cylinder or truncated cone, the working surface in this case is the side with the smallest area S_c (figure 1). Section of the cathode opposite to its working surface is cooled with running water. Continuous chaotic movement of cathode spots at a fixed surface of the cathode [6], regardless of the initial distribution, leads to some average equilibrium temperature of the surface, determined by the geometrical dimensions of the cathode: the diameter and length, and conditions of its cooling.

Diameter of the working surface of the cathode may be as much as 200 mm. Geometric dimensions are selected to provide lowest average operating temperature of the cathode. During the work of the source due to erosion of the material the length of the cathode decreases (figure 1) that leads to a decrease in temperature of the cathode and to changes in the conditions of plasma flow formation.

Thus, the temperature of the cathode is the main factor determining modes of generation of the plasma flux, and therefore, even providing the stabilization of the discharge parameters, as the cathode is sputtered, there are significant changes in it leading to deviations of thickness of the formed coatings. In batch deposition of coatings these conditions require adjustments to the process.

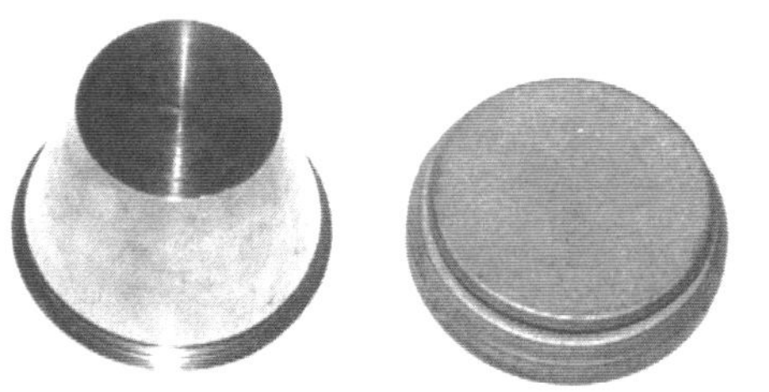

Figure 1. Consumable cathode of a vacuum arc plasma source before (left) and after (right) work cycle.

If the surface receives a flux of ions with energy W_i and the ion current density j_i, from the condition of mass balance, the rate of coating growth is determined by the expression [7]:

$$\upsilon_{\text{dep}} = j_i(z)\left(\chi(W_i) - S(W_i)\right) / \xi e n_0, \quad (1)$$

where $j_i(z)$ – is the density of ion flow to workpiece; $\chi(W_i)$ – accommodation coefficient (the fraction of the condensed ions); $S(W_i)$ – sputtering rate (fraction of ions leaving the surface layer); ξ – average ion charge in plasma; n_0 – concentration of particles in surface, uncompacted layer with a characteristic thickness, equal to the path of ions.

Average in time and space magnitude of the ion current density at the cathode axis and at a distance z from it is possible to determine, based on the assumption that the supersonic flow of plasma outbound from the cathode area extends in the radial direction with the speed of ion sound:

$$j_i(z) = 4I_i / \pi\left(D_c + 2z\frac{\upsilon_{is}}{\upsilon_{\text{pl}}}\right)^2,$$

where $I_i = 0.08 I_{\text{dis}}$ – value of the ion current; D_c – diameter of the cathode.

Coefficients χ and S included in the equation (1) depend on the energy of ions bombarding the surface W_i, which consists of the initial energy $W_{i0} = m_i \upsilon_{\text{pl}}^2 / 2$ and the energy acquired in Debye layer $e\xi U_{\text{bias}}$, adjacent to the substrate, provided that it is set to the accelerating negative potential U_{bias}. Changing the value of U_{bias} allows to adjust the energy of the ions, and, consequently, to control the flow of the process, turning it from a spray mode to the formation of the coating. Thus, the process of coating formation is determined by the energy of the condensing particles, which influence not only on the adhesion of the coating, but also on its structure, composition and presence of defects.

Figure 2 shows dependence of the growth rate of the applied coating from the time of the evaporator operation. The original length of the titanium cathode was equal to $4.5 \cdot 10^{-2}$ m, with a diameter $6.0 \cdot 10^{-2}$ m. Change in the length of the cathode in the process corresponds to a change in its operating temperature in time. Transition of the dependence into the horizontal section coincides with the transition of the temperature for the studied material, through the point of the recrystallization annealing. Cathode spots on the working surface of a titanium cathode at a higher temperature leave clearly delineated melted zones with a depth of several millimeters. At a lower temperature the working surface of the cathode has a fine-grained structure with a metallic luster.

With increasing operating temperature of the cathode in the flow increases the percentage of neutral vapor and droplet formation [8], which is confirmed by the dependency of the growth rate of the applied coating from the negative bias, set on the substrate, obtained when the length of the cathode is $4 \cdot 10^{-2}$ and $2 \cdot 10^{-2}$ m (figure 3) and respectively at different temperature.

When the length of the cathode is $4 \cdot 10^{-2}$ m (curve *1* in figure 3) value of the bias voltage has no effect on the growth rate of the formed coating, since the flow is dominated by the presence of droplet formation. During sputtering of the cathode material, a decrease in its working temperature is observed. At the end, when the length of the cathode decreases, all the power that is released at the cathode is withdrawn by the cooling system. Under these conditions the growth rate is determined by the positively charged component of the plasma flow (dependence *2* in figure 3, length of the cathode is $2 \cdot 10^{-2}$ m). The observed decrease in coating thickness with increase in negative bias is associated with the effect of sputtering caused by bombardment of the surface by high energy ions.

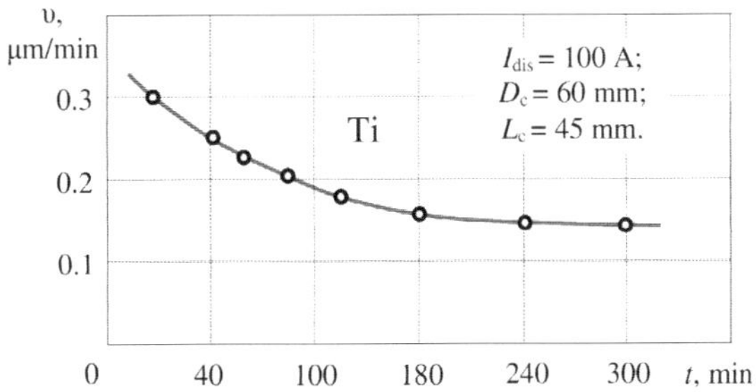

Figure 2. Dependence of the growth rate from the time of the evaporator operation.

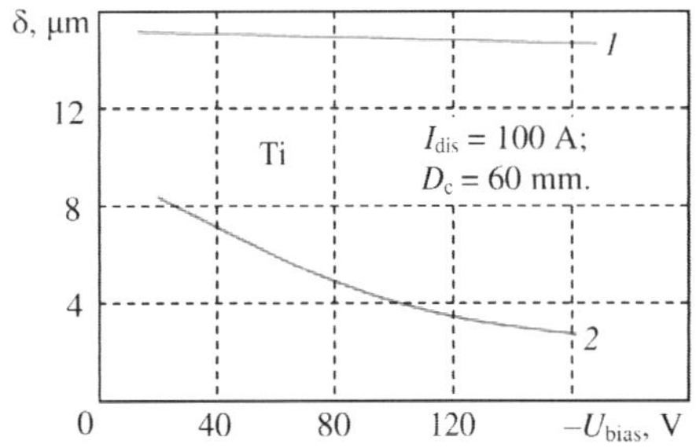

Figure 3. Dependence of the growth rate from the negative bias.

Performed using X-ray diffractometer DRON-3 in CuK_α radiation metallographic investigations showed that the temperature of the cathode depends on the structure of the formed coating. For titanium, at a temperature of the cathode below 880 °C obtained coating had a complex structure. The increase in the magnitude of the negative bias (in absolute value) on the treated substrate leads to the formation of coatings with very distinct components (111), (110) and (012), while at the temperature of the cathode in excess of 880 °C the effect of the orientation was weak and the coating had a loose, rough character. Thus, the temperature of the growing film and the phase composition of the flow have a crucial impact on the properties of the obtained coatings. In this regard, the main problem when using vacuum arc plasma sources is the operational control of the thickness of the formed coating.

All methods of measuring coating thickness can be divided into two large groups based on destructive and nondestructive control methods. In each case the choice of method and equipment depends on the parameters of the controlled sample. To destructive control methods, in particular, may be referred a method of determining coating thickness using a spectrophotometer "Saturn", the principle of which is based on the conversion of the sample into atomic state, with subsequent photometric transformation of the optical density of the atomic vapor into an electrical signal. The thickness of the coating in this case can be determined from the following relationship:

$$\delta = m_s / \rho ,$$

where m_s – is the specific surface density of the coating; ρ – bulk density of the coating material.

Information about the concentration of the analyte is obtained as a result of absorption by free atoms of radiation from a spectral source – lamp LSP-1. Test and control samples are processed, whereby the mass is determined using a preset concentration in the control samples. The RMS relative deviation of this method for titanium does not exceed 10 %.

Of special interest are methods of nondestructive testing [9]. Coating thickness can be determined using control sample, which has a step between the film and the substrate (scratch on the surface of the substrate or edge of the film on the open substrate). According to the obtained topography, the thickness is measured with great precision using profilometric methods. Upon reflection from the surfaces of the film and substrate, assuming the existence of the relief are applicable interferometric measurements using microinterferometers MII-4.0, MII-11 or MII-12, and to a certain extent, these measurements can be automated. The resistive method allows estimating the film thickness by its ohmic resistance. To implement this method in a thread, along with the substrates is placed a control dielectric plate, rectangular in shape, which allows applying when of a well-known expression for the resistance of a conductor with a given length and a rectangular cross section.

Capacitive measurement method is only applicable for dielectric films and coatings, and impossible for the metal and semiconductor layers. In addition, this method gives the average value of the coating thickness across the object. Also only for dielectric and semiconductor films can be applied becoming increasingly common spectrometric measurement method [10, 11].

Wide application, due to the high sensitivity (to the deposited mass it is ($10^{-8}...10^{-11}$) g·cm^{-2}·Hz^{-1}) and the ability to control the weight, thickness and speed of material application, has a method of quartz resonator. However, its use is hampered due to the temperature drift of frequency of the quartz resonator, which requires additional cooling [7]. Ultrasonic method is applied to coatings with a thickness more than 0.1 mm, the measurement error is 5 %. But the application of this method is limited because of the strong attenuation of ultrasonic waves in thick coating layers, which complicates acquiring of a weak reflected signal due to complex shapes of objects. In addition, when ultrasonic method is used in some cases there can be a detachment of the coating.

The vortex method is based on the analysis of the interaction of electromagnetic field with the electromagnetic field of eddy currents induced by the exciting coil in an electrically conducting test object. Apparatus that implements this method allows to measure thickness from a few micrometers to millimeters, with an accuracy of 2...5 %. The minimum area of the control zone can be increased to 1 mm^2, which opens the possibility of measuring coating thickness on complex objects. However, this method requires the combination of certain material properties of the coating and the substrate.

All methods of coatings thickness measurement considered above have limitations for practical application. To determine the thickness of the coating from units to tens of microns, provided control efficiency and versatility of the investigated coatings were selected methods based on a quantitative X-ray phase analysis. The coating thickness measurement was performed using diffractometer ADP-2M in two ways relating to the registration of the diffraction pattern of the coating or substrate.

In the first case, the reflected rays leave the crystal from the side of a primary beam and are recorded in front of the crystal. Thus is examined the change in the intensity of the selected diffraction reflection of a coating depending on its thickness. In this case, the calculation of a coating thickness is provided by the following equation [12]:

$$I_{HKL} = I_0 QS\,[1-\exp(-\gamma h/\sin\vartheta)]/2\gamma\,, \tag{2}$$

where I_{HKL} – is the intensity of a diffraction reflection with indices H, K, L; I_0 – intensity of the primary beam; Q – integral reflectivity of the unit; S – area of the cross section of the beam; γ – linear attenuation coefficient of the used characteristic X-ray radiation by the coating material; ϑ – angle of diffraction; h – thickness of the coating.

When conducting the experiment in such way the area of the irradiated surface and the acquisition conditions of the diffraction pattern remain unchanged, thus it is possible to associate the intensity change with changing layer thickness. For a sufficiently thick cover, when $h \to \infty$ the equation (2) takes the following form:

$$I_{HKL}^{M} = I_0 QS/2\gamma\,.$$

Table 1 shows how the ratio of the intensities of some diffraction reflections of titanium film and bulk samples is changed during registration in CuK_α radiation.

Table 1. Ratio of the intensities of some diffraction reflections of titanium film and bulk samples I_{HKL}/I_{HKL}^{M} during registration in CuK_α radiation.

Reflection indexes *HKL*	ϑ, °	Coating thickness, μm					
		0.5	1.0	3.0	5.0	10.0	15.0
(011)	40.2	0.24	0.42	0.80	0.93	0.99	1.00
(021)	77.4	0.14	0.26	0.60	0.76	0.95	0.99
(123)	139.7	0.09	0.16	0.40	0.58	0.82	0.93

Thus, choosing the appropriate reflection for a pattern gives the opportunity to efficiently measure coating thickness in the range from 0.5 to 15 μm. Changing the wavelength of the used characteristic radiation from CrK_α to MoK_α extends the limits for determining thickness by this method to 40 μm.

In the case of thin substrates measurement of coating thickness is possible by the weakening of the direct X-ray beam. The attenuation of the radiation intensity is according to the following law:

$$I_h = I_0 \exp(-\gamma\, h)\,.$$

Since the attenuation coefficient depends on the wavelength of X-ray radiation it is necessary to use a tube with small energy quanta of radiation. Thus, for this analysis it is necessary to have the X-ray tube with a possible lowest energy of a characteristic radiation. The study of samples was carried out as follows. The sample is placed on the way of direct X-ray radiation. First the intensity of X-ray radiation of the sample without coating is determined and then the measurements are repeated for a coated sample. The obtained data is used to calculate the thickness of the investigated coating according to the following equation:

$$I_h / I_0 = \exp(-\gamma_1 h_1)\exp(-\gamma_2 h_2),$$

where γ_1 – is the linear attenuation coefficient of the used characteristic X-ray radiation by the material of the coating with thickness h_1; γ_2 – linear attenuation coefficient of the used characteristic X-ray radiation by the substrate material with thickness h_2.

The results of measurement of the coating thickness for various samples using a method, based on the X-ray attenuation, are presented in table 2. As the substrate was used aluminum material and copper as a coating. Parallel control of the coating thickness was carried out by weighting method and the study of metallographic sections.

Table 2. Results of the coating thickness measurements for various samples using a method, based on the X-ray attenuation.

Measurement method	Coating thickness, μm				
Weighting	0.78	1.10	1.30	2.10	2.40
X-ray attenuation	0.80	1.10	1.10	1.90	2.10

As follows from the table 2 the proposed X-ray method of measuring the coating thickness gives a satisfactory agreement of data with relatively reliable measurement using weighing of the samples.

Using results of this work were optimized the parameters of a vacuum arc metal plasma source with coaxial design of the electrodes. Also was showed the possibility of rapid measurement of the thickness of the applied coating with a sufficient degree of accuracy by the X-ray methods.

References

[1] Kostrin D K and Lisenkov A A 2016 *Materials Science Forum* **843** 278–83
[2] Kostrin D K and Lisenkov A A 2016 *Materials Science Forum* **870** 371–6
[3] Vetrov N Z, Kostrin D K, Lisenkov A A and Popova M S 2015 *J. Phys.: Conf. Ser.* **652** 012032.
[4] Vinogradov M L, Kostrin D K, Smelova V V, Trifonov S A and Lisenkov A A 2016 *Proceedings of the 2016 IEEE North West Russia Section Young Researchers in Electrical and Electronic Engineering Conference* 729–30
[5] Bystrov Yu A, Vetrov N Z, Lisenkov A A and Kostrin D K 2014 *Vakuum in Forschung und Praxis* **5** 19–23
[6] Bystrov Yu A, Kostrin D K, Lisenkov A A and Vetrov N Z 2015 *Vakuum in Forschung und Praxis* **2** 22–5
[7] Bystrov Yu A, Kolgin E A and Kotletsov B N 1988 *Technological control of dimensions in microelectronic production* (Moscow: Radio i sviaz)
[8] Lisenkov A A and Vetrov N Z 2000 *Vacuum arc plasma sources* (Saint Petersburg: Energoatomizdat)
[9] Martsinukov S A, Kostrin D K, Chernigovskiy V V and Lisenkov A A 2016 *J. Phys.: Conf. Ser.* **735** 012049
[10] Kostrin D K, Uhov A A and Lisenkov A A 2016 *Vakuum in Forschung und Praxis* **3** 34–7
[11] Kostrin D K, Lisenkov A A and Uhov A A 2016 *J. Phys.: Conf. Ser.* **735** 012055
[12] Ivanov S A, Komiak N I and Pavluk E G 1979 *X-ray methods of substances crystal structure research* (Leningrad: Leningrad Electrotechnical institute)

IOP Conf. Series: Journal of Physics: Conf. Series **808** (2017) 012006 doi:10.1088/1742-6596/808/1/012006

Synthesis of carbon nanotubes on the surface of the field cathode of X-ray tube

M L Taubin[1,2], A A Urusov[1] and A V Bratsuk[1]

[1] Federal State Unitarian Enterprise Scientific Research Institute the Scientific Industrial Association "Luch", 142100, Moscow region, Russia
[2] National Research Nuclear University "MEPhI", 115409, Moscow region, Russia

E-mail: taubin@sialuch.ru, taubin41@rambler.ru

Abstract. Carbon nanotubes are unique materials being actively investigated in the development of X-ray tubes with a field cathode. The so-called cold-ray tubes. Our contribution in this direction is the use of single crystals W for the anodes and cathodes as substrates. Here are the structures we obtained carbon nanotubes with the catalysts and directly on the surface of tungsten. The report presents results of studies of carbon nanotubes (CNT) formation on metal substrates W and Ni. X-ray microprobe analysis revealed the presence of 8 at. % carbon. Average density of the powder was measured as 5.1 ± 0.1 g/cm^3. Calculations have shown that this value corresponds to the density of the carbon content of 8 at. %.

1. Introduction

The aim of this research was studying the possibility of coating the surface of tungsten by carbon nanotubes (CNTs). In case of success it would appear the experimental confirmation of the possibility of formation of CNTs on a tungsten substrate as a field emission cathode of X-ray tubes [1–3].

2. Methods and results

Most of the samples used in the experiments for growing nanotubes were thin metal plates sizes of 10×10 mm mainly made from nickel and tungsten. Before the process of applying CNTs surface samples polished to obtain a developed surface and etched in acid to remove the oxide layer. The liquid catalyst in the form of solution of nickelous nitrate [$Ni(NO_3)_2$] was applied on some samples from a tungsten.

The prepared samples were placed into a quartz cuvette, which was placed into the reactor for the synthesis of nanotubes, a circuit diagram is shown in figure 1. Then include heating the reactor chamber at a predetermined rate to a predetermined temperature are controlled thermostatic device. During the heating in the volume of the chamber is injected inert gas argon. Upon reaching required temperature included a heating device of the evaporator and begin feeding in the chamber with sample a pair of ethanol. The process was stopped after a specified time or after the complete evaporation of the ethanol.

In work as a carbonaceous reagent used ethanol, which has a number of advantages compared to other the reagents (e.g. propylene). Under normal conditions the ethanol exists in a liquid state and is easily stored. It is cheap and low toxic, have no high boiling point (78 °C) and decomposition temperature (about 600 °C). From the chemical standpoint feature of ethanol consists that at pyrolysis of its vapor OH-radicals are formed, which have high reactivity with respect to defective (amorphous) carbon, due to that the CNTs are more pure. Also the OH-radical prevents contamination of the catalyst particles by amorphous carbon which increases time of growth of nanotubes and as a consequence increases the length.

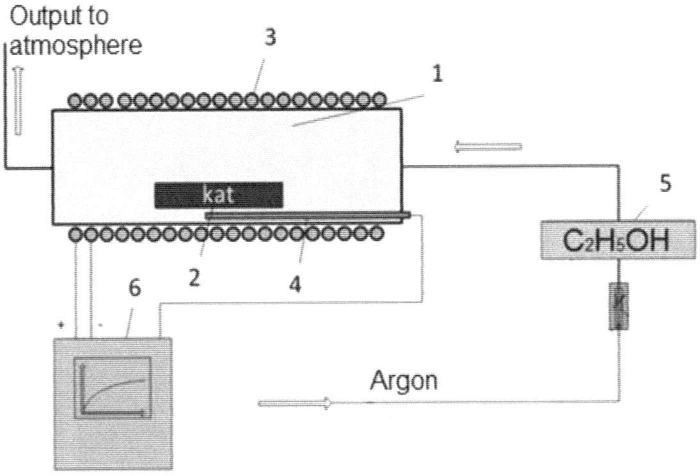

Figure 1. The scheme of the reactor for the synthesis of CNTs: 1 – quartz chamber; 2 – cuvette with the sample; 3 – heating element; 4 – thermocouple; 5 – evaporator with ethanol; 6 – temperature-control systems.

The experimental results (figure 2) showed that on a surface of a sample needle phases were created, and in some places crystalline structures of the faceted form were formed.

Element	Mass., %
C	11.01
O	27.20
W	59.20

Figure 2. The image of the tungsten surface of substrate and a local elementary analysis of this area.

In connection with the foregoing a need to use catalysts, particularly nickel. Figure 3 shows the structure of CNTs on a nickel substrate. These results confirmed prospects of use of nickel as catalyst material. In subsequent experiments, for example, by controlling the thickness of the applied nickel layer can be controlled by the size and number of synthesized nanotubes.

In preparing the tungsten substrates, solution of nickelous nitrate [$Ni(NO_3)_2$] was applied on their surface and then process of synthesis of CNTs was carried out. The results are shown in figure 4.

IOP Conf. Series: Journal of Physics: Conf. Series **808** (2017) 012006 doi:10.1088/1742-6596/808/1/012006

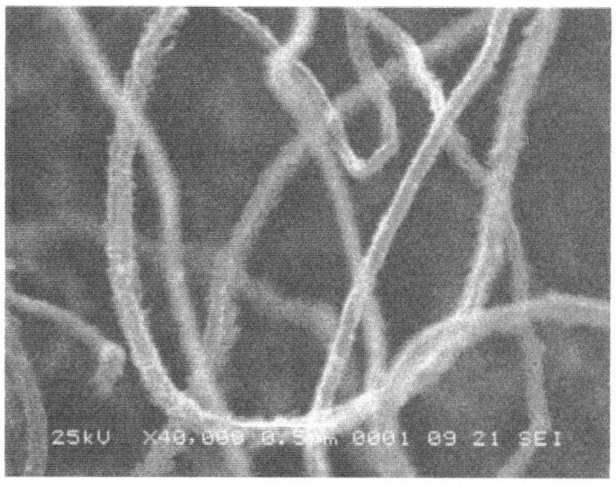

Figure 3. Electron microscopic image CNTs on a nickel substrate.

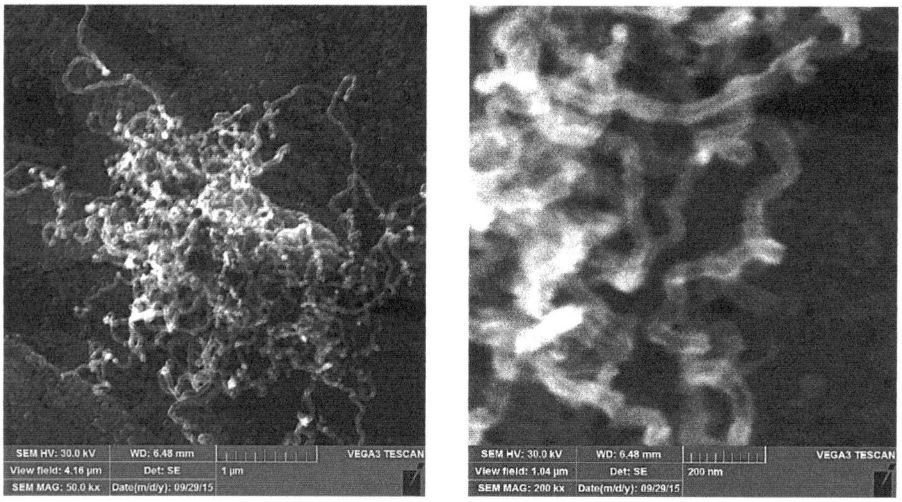

Figure 4. Images of a surface of a tungsten substrate pre-coated $Ni(NO_3)_2$ at increase in 50000 (left) and 200000 (right) times respectively.

Based on the results of electron microscopic researches been established that the presence of nickel catalyst leads to stable formation of CNTs on the tungsten surface.

3. Discussion of research results

In the first experiments the modes of synthesis of carbon nanotubes on nickel substrates with ethanol use as carbon source for definition of dependence of process parameters on nanotubes structure such as temperature, time and feed rate of gaseous components were fulfilled. The established laws of formation carbon structures at various parameters are the result of experiments.

Application of carbon nanotubes on substrates from tungsten for receiving nanotubes in the production technology of cathodes for X-ray tubes was the following stage of researches. The nickel catalyst which was applied on substrates from tungsten for formation on their surface of the CNTs arrays was used at the subsequent stage [2].

The thermal tests which showed thermal stability of carbon nanotubes to $T = 1700$ °C were carried out after receiving positive takes of the above researches. Additionally, by scanning electron microscopy radius nanotubes produced was measured, which ranged from 20 to 100 nm. The value of the specific surface was determined by the method of Brunauer–Emmett–Teller, it was 92.64 m^2/g. The modes of getting CNTs are worked, and also influence of process parameters of their getting on structure of the nanotubes array on the tungsten surfaces is established.

4. Conclusions

It was established that synthesis of carbon nanotubes by method of a catalytic pyrolysis immediately on the surface of tungsten with ethanol use as carbon source at the chosen modes did not lead to synthesis of CNTs. However, the application of a nickel catalyst on the substrate surface tungsten gave positive results in experiments on synthesis of CNTs on the tungsten surface.

References

[1] Yeletsky A V 2002 *Progress physical sciences* **4** 401–38
[2] Yeletsky A V 2010 *Progress physical sciences* **9** 897–930
[3] Harris P 2003 *Carbon nanotubes and congeners structures. New materials of the 21st century* (Moscow: Technosphere)

Measuring equipment for controlling the anode current during training and testing of the X-ray tubes

N O Perez Vasquez, D K Kostrin and A A Uhov

Department of electronic instruments and devices, Saint Petersburg Electrotechnical University "LETI", 197376, Saint Petersburg, Russia

E-mail: dkkostrin@mail.ru

Abstract. In this paper the features of controlling of the anode current of the X-ray tubes during their training and testing are described. Requirements for developing of the control and measuring equipment are given. In detail are shown the approaches to the development of two types of microammeters with optical signal transmission: analog, with compensation of the nonlinearity of the transfer characteristics with the use of the optical negative feedback, and digital, in which the values of anodic current are transfered using a microcontroller.

1. Introduction

Training and testing of the X-ray tubes are very important stages of their production, and allow achieving a significant increase in the quality of manufactured products [1]. The main aims and objectives of the training of the X-ray tubes are:

- electric polishing of the electrodes surfaces due to the evaporation of micro edges (enhancement of the electric strength);
- cleaning of the surfaces by electron-ion bombardment from the products of the cathode evaporation (reduction of the leakage current);
- removal of gas atoms from the surface layers of the electrodes and their absorption by the getter (stabilization of the vacuum condition during operation).

During training of the X-ray tubes it is necessary to control the current value in the X-ray tube circuit from a few μA to tens of mA at a potential of 160 kV with respect to the ground. Training is conducted in the two consecutive stages. During the first phase a high voltage is applied to the X-ray tube with a cold cathode. With the control of the leakage current the training continues for as long as this current is not reduced or at least becomes stable. The current at this stage may be in the range of 1...50 μA. Supply voltage can be accompanied by the electric breakdowns that need to be counted.

In the second phase, training is conducted with a gradual increase in the filament current, and consequently the anode current of the X-ray tube. Nominal anode current usually does not exceed 50 mA, however, during breakdowns a measurement device must be able to withstand current up to several amperes. Thus, the scale of measurement of the microammeter should be no less than 5 decades. The device should count the number of discharges occurring in the X-ray tube at both stages of training and maintain its performance when exposed to X-ray radiation up to 200 keV.

For many years for measuring current and voltage in the test process units pointer instruments were used (figure 1), however, this measurement method is only possible for static tests. Generally, measuring instruments are placed inside the test chamber where they are exposed to strong electromagnetic fields and X-rays. The installation of the necessary current values or its measurement is associated to the operator with certain difficulties, since it is necessary to observe the readings at a considerable distance from them through a protective window made of lead glass.

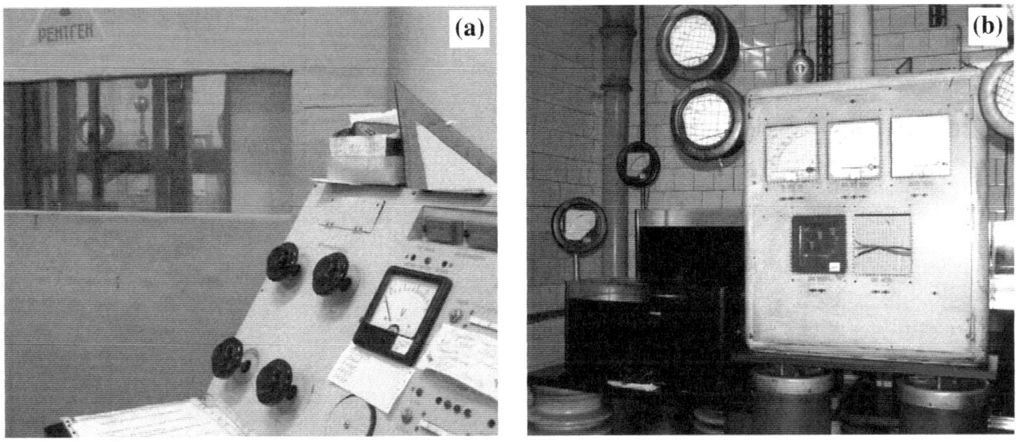

Figure 1. Control panel (operator's workstation) (**a**) and the location of the pointer-type ammeters in the training chamber (**b**).

As was mentioned earlier, during training and testing of the X-ray tubes, both the leakage currents of a few microamperes and operating currents, reaching tens of milliamps, are subject to control. This fact forces to install into the high-voltage circuit multiple meters with different measuring ranges, each of which is connected to the measuring circuit with a relay, insulated for a full voltage.

The problem is further complicated if tests are carried out for the pulsed X-ray tubes. The duration of the current pulses in these devices can vary from a few hundredths of second to a few seconds. Even the most modern analog ammeters do not permit the necessary measurements. Meanwhile, the production of the pulsed X-ray tubes has been increasing steadily. The advantages of pulse mode are undeniable both from the standpoint of radiation safety for personnel and the metrology characteristics of medical and translucent X-ray devices.

Based on the X-ray production situation, the task of developing new instruments to measure direct and pulse current in the high-voltage circuits arose. It is necessary to seek a solution based on the fact that modern technologies require information in a digital form. This opens up opportunities for processing and documentation of measurement results and creates more favorable conditions for the work of the operator [2].

2. Development of the control and measurement equipment

The current in the high-voltage circuit can be controlled by measuring the voltage drop across the shunt, which is connected directly into the high-voltage circuit, with the subsequent transfer of the signal proportional to the current to the operator panel. The conversion of the voltage drop on the shunt into the signal transmitted to the remote operator can be implemented by means of the conventional analog and digital circuits under the high-voltage circuit potential, using an independent low-voltage power source. There can be distinguished three different types of transmission channels that provide a reliable electrical insulation at a voltage of X-ray devices in the hundreds of kilovolts:

- radio channel;
- direct optical channel;
- fiber optic channel.

Data transmission over the radio channel provides a number of significant advantages; however, in the conditions of the test plant, when operating high-voltage X-ray units, intensive noise is inevitable in almost all the wavelengths range, the result is a huge technical challenge to provide a reliable radio channel. The connection between the two units of the ammeter can be carried out by a direct optical channel, for example, in the infrared range. However, a special chamber, which houses the X-ray device, has only a small window of leaded glass, which greatly complicates the solution of the problem in this way.

Let us consider fiber optic as a communication means for transmitting data from the high-voltage unit to the operator panel. Fiber optic cable can be easily output from a high-voltage chamber, in which the X-ray device is placed. Usually a cable length of 10 m is sufficient; with such a small length damping in fiber optic communication lines will be very small – if a loss factor is taken equal to 10 dB/km, the intensity of the light flux at the output of the line is only 1.2 % less than at the entrance. Additionally, the signal propagating through the fiber is not susceptible to electromagnetic interference, which is very important in the operation of the device in proximity to X-ray installations.

The simplest converter of the current into an optical signal with a subsequent inverse transformation of the optical signal into the current or voltage is a volstron – device in which the emitter and the photodetector are nondetachably connected to each other with a fiber optic cable. The simplest volstron optocoupler consists of a light-emitting diode and a photodiode, separated by an optical fiber and has a significant nonlinearity of the transfer characteristics. One of the most effective and meanwhile simple means of linearizing the transfer characteristics of the volstron is the introduction of a negative feedback circuit.

To create thermostable volstron should be used such light-emitting diodes, the energy characteristics of which are minimally dependent on temperature. The second and main component of the optical system is the photodetector. After selecting the light-emitting diode it is required to coordinate its emission spectrum with the spectral sensitivity of the photodetector. The third component – fiber optic line (including optical connectors) must provide signal transmission with minimal losses and have high stability of parameters.

On the basis of selection of the optical elements, it is possible to proceed directly to creating a circuit of linear converter of the voltage into the proportional optical signal. It is known that the power to the light-emitting diode from the output of the amplifier covered by the feedback through the optical channel using the photodiode operating in the current source regime, allows to convert an input voltage into the proportional radiation flux with an error of a fraction of a percent and keep its temperature stability, provided that the optical components are stably optically aligned. This principle is used in the presented circuit of the voltage-to-voltage converter with optical isolation (figure 2).

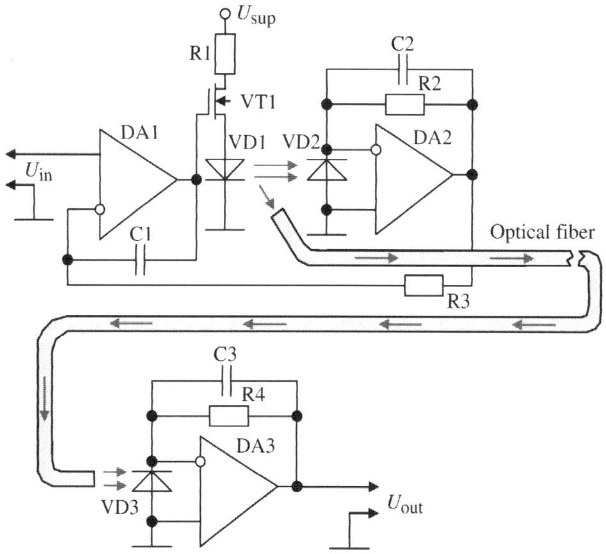

Figure 2. Linear voltage-to-voltage converter with optical isolation.

First, the input voltage U_{in} is converted into a current of the light-emitting diode VD1. Used in the circuit of the voltage-to-voltage converter a low-power operational amplifier has insufficient current output for direct control of VD1. To increase the current a field effect transistor VT1 is additionally applied. Feedback at the operational amplifier DA1 is supplied via resistor R3 from the operational amplifier DA2, which amplifies the signal from the photodiode VD2.

Thus, the input voltage is linearly transformed into a luminous flux generated by the light-emitting diode VD1. Part of the VD1 radiation is supplied via optical fiber to the receiving photodiode VD3. Since the photoelectric converter DA3 has a high linearity of conversion in not less than 4–5 decades, the output voltage U_{out} is linearly dependent on the input voltage, proportional to the measured current. Output voltage from the VD3 is passed through another amplifier to the microcontroller where it is processed and then the digital measurement result is displayed.

Although linearity of volstron converter fully meets the needs of production, its dynamical range does not exceed three decades. When training some X-ray tubes it is necessary to control the values of current from a few μA to tens of mA. Thus, the scale of measurement of the microammeter should be no less than 5 decades. Measurement of currents in units of microamps requires using at the input of transducer a precision operational amplifier with automatic zero point correction and offset voltage not more than units of microvolts. The task of transmitting signals proportional to the current from the measuring shunt to the input of the conversion and information processing device over a wide dynamic range from 0.5 mA to 50 mA, requires placement of the devices for conversion of the current into a digital code under the high voltage.

To convert the voltage drop at the shunt in the current circuit into a digital code and transmit it according to a specific protocol to an operator's console an analog-to-digital converter and a microcontroller must be placed in the high-voltage unit (figure 3).

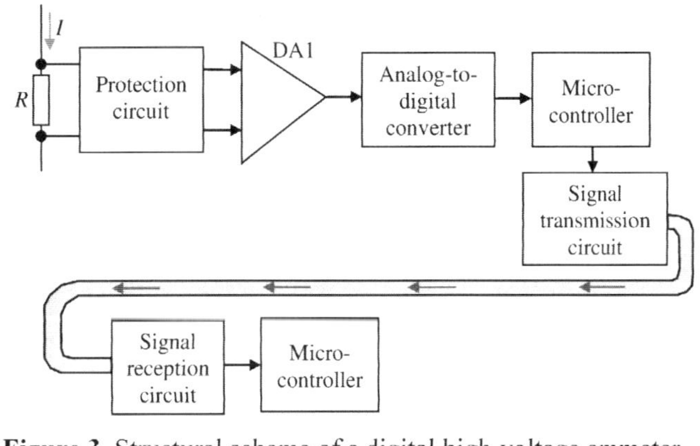

Figure 3. Structural scheme of a digital high-voltage ammeter.

The advantages of measuring current using such high-voltage blocks are obvious:

- the use of multi-bit analog-to-digital converters with internal programmable gain allows measurement over a very wide dynamic range – up to 5 orders of magnitude;
- the presence of a microcontroller makes it possible to transfer data using any protocol that allows to calculate and send a data checksum that virtually eliminates the possibility of undetected errors in the transmission.

At the same time, such device would also possess significant drawbacks:

- high energy consumption – in addition to powering up the microcontroller and analog-to-digital converter even the transmission of a zero level data consumes energy. Considering that the devices must be supplied from batteries, this will significantly reduce the working time;
- in the absence of grounding powerful electromagnetic interference when operating X-ray equipment may adversely affect both the conversion accuracy of the analog-to-digital converter and the operation of the microcontroller;

- there is a possibility of loss of communication with the conversion device under irradiation of a high flux X-ray radiation.

Microchips of ultrahigh degree of integration have a high sensitivity to exposure to X-rays and for this reason the developed device requires the use of a protective case that reduces radiation exposure of a measuring circuit to a safe level.

The current converter must operate in the conditions where depending on the type of the X-ray tube, the maximum energy of quantums does not exceed 320 keV. Protection may represent a screen of iron and lead, which is simultaneously the housing of the device. The case is a metal shell of stainless steel with a wall thickness of 0.5 mm, in which the batteries and a lead box containing a printed board with microchips are placed. The box is made of lead plates 5 mm thick due to the calculation results. Figure 4 shows the measuring blocks of two digital microammeters placed in the test chamber in close proximity to X-ray tubes.

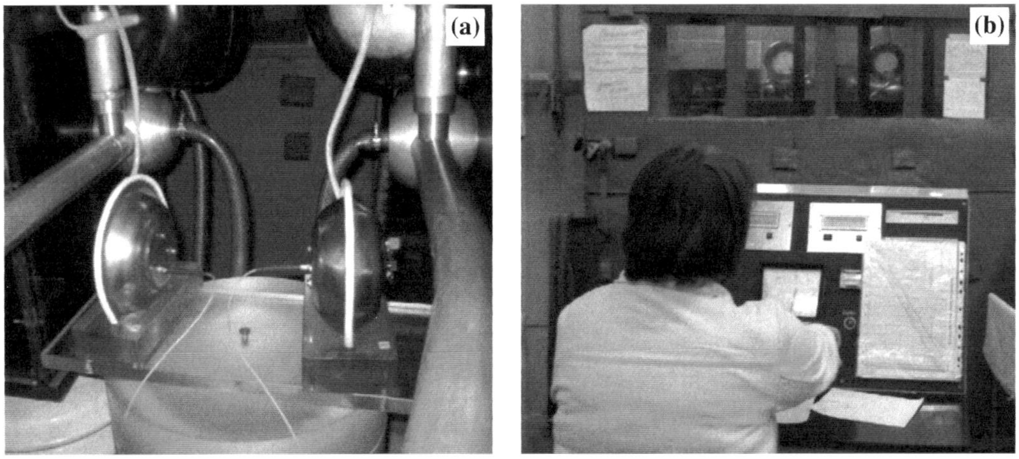

Figure 4. Measuring units of the digital microammeters in the test chamber (**a**) and the information processing units on the operator panel (**b**).

With the help of the developed device operator during the training and testing can detect the change in the leakage current and determine how many breakdowns in the X-ray tube occurs at certain stages of the training. This information allows making a conclusion on whether to send the X-ray tube for additional pumping or not. Ultimately, the application of the developed device allows improving the quality of the products. At the end of training depending on the results of measurement of the leakage current and the number of breakdowns, it is possible to classify the X-ray tube according to quality categories and make a prediction of the possible service life.

3. Conclusions

Using the solutions proposed in this work have been manufactured and delivered to the company two types of experimental models of electronic microammeters for the divisions of training and testing of the X-ray tubes, which are already several years successfully used in workshop conditions.

The developed equipment is capable of measuring pulsed and continuous anode current in the circuits, located under the high voltage, with wide dynamic range and leakage current of the X-ray tube, simultaneously counting the number of electric breakdowns and showing information to the operator in a digital form.

References

[1] Bystrov Yu A, Kostrin D K, Peres Waskes N O and Ukhov A A 2013 *Biotechnosphere* **4** 60–4
[2] Vasilevskii A M, Konoplev G A, Svetlov D A, Gerasimov V A, Kostrin D K and Ukhov A A 2014 *Biomedical Engineering* **1** 13–6

Cutting-edge technologies of medical radiography: tomosynthesis and double energy

A O Ustinov[1] and A R Dabagov[2]

[1] LLC Scientific-technical center "MT" (STC-MT LLC), 143026, Moscow, Skolkovo, Russia
[2] CJSC Medical Technologies Ltd (MTL CJSC), 105318, Moscow, Russia

E-mail: uao@mtl.ru, dar@mtl.ru

Abstract. The tomosynthesis technology is on the fringe of linear and computer tomography, combining high information capacity and low exposure. Diagnosing pathology of chest organs at early stages is complicated due to overlapping views of the lung and bone tissues. By using different X-ray energy ranges tissues with different absorption factors are separated and their separate images are received. This technology ensures that pathology is revealed at the early stage during routine screening.

Modern medical radiography can fulfill most clinical diagnostic tasks. Thanks to the flat-panel digital X-ray detector technology, the patient's exposure can be reduced significantly as compared to the traditional analog radiography. Traditional X-ray diagnostic methods have reached their limits. To increase medical practice efficiency further, we have to switch to using additional visualization methods, which include tomosynthesis and double-energy technologies.

In general-purpose radiography, the traditional linear tomography yields layer-by-layer images of the object, but only one slice is obtained during one pass of the X-ray generator. As a result, if a layer-by-layer image of the entire lungs is required the procedure is performed at least 10 times, which results in a considerable increase of the patient's radiation exposure and the procedure duration. The computer tomography method provides information about the entire chest volume, but increases the patient's radiation exposure significantly [1].

The tomosynthesis technology is on the fringe of linear and computer tomography (CT), combining high information capacity (as compared to linear tomography) and low exposure (as compared to computer tomography). The X-ray source travels around the region of interest in an arc, making a series of exposures. The data on the geometric location of the source and the corresponding angled views make it possible to reconstruct the patient's internal volume.

There are a large number of tomosynthesis reconstruction algorithms:

- shift-and-add method (SAA);
- filtered back projection method (FBP);
- maximum likelihood estimation method (MLEM);
- adaptive statistical iterative reconstruction method (ASiR).

Each algorithm has its own advantages and shortcomings. SAA is one of the simplest algorithms, but the reconstruction quality leaves much to be desired. The algorithm adds views up layer by layer with a certain shift providing for the required height of the reconstructed layer. The FBP algorithm is the most popular one as of today, as it ensures the required image quality and is not resource-intensive. The main peculiarity of this algorithm is the choice of the filtering function. Its choice affects the image quality and presence of artifacts greatly. Iterative algorithms, MLEM and ASiR, are highly resource-intensive, which prevents their wide use in commercial systems [2].

Examples of reconstructing a calf's head and the thorax of a Kyoto Kagaku PBU70 pediatric phantom using the FBP algorithm are presented below (figures 1, 2). The reconstructions were made using the Telekord-MT-Plus device made by CJSC Medical Technologies Ltd with the angle of 60° and in 30 views.

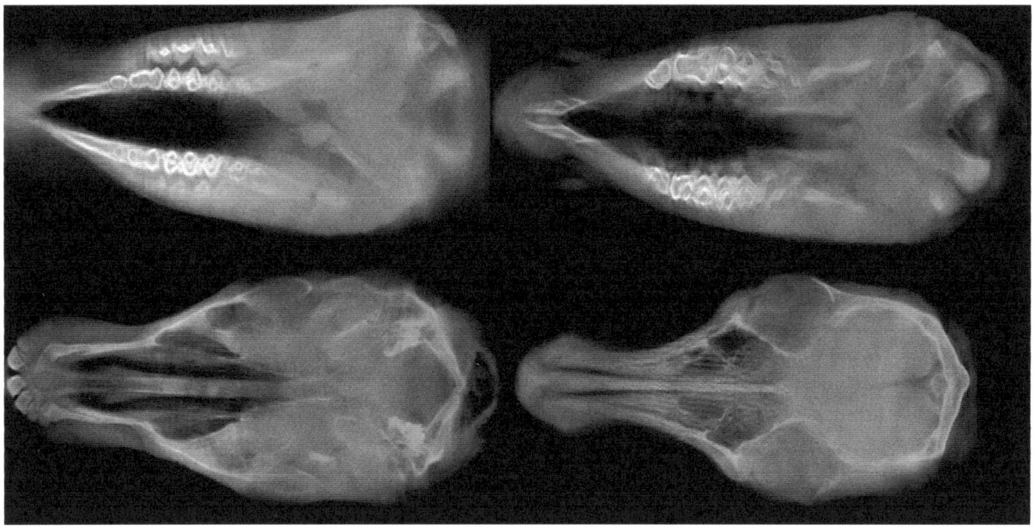

Figure 1. Sample calf's head reconstruction.

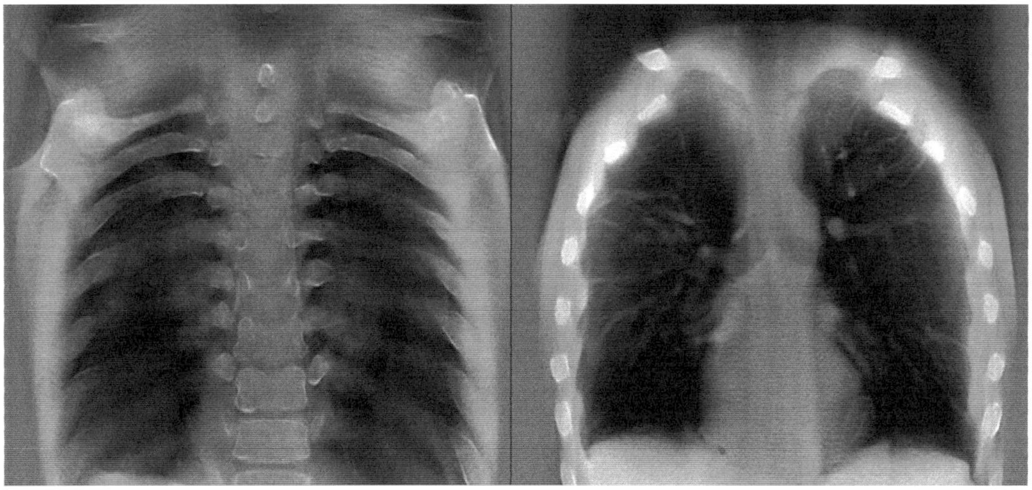

Figure 2. Sample thorax reconstruction.

The Telekord-MT-Plus device allows conducting tomosynthesis studies with the column angulation angle and with the number of views up to 60. The operator sets the height of the region of interest in centimeters in accordance with the centerline of the object studied or the assumed location of the lesion. Besides, limit values for the reconstruction scope height and the slice thickness in millimeters need to be set. This ensures a high level of detail of the images received.

IOP Conf. Series: Journal of Physics: Conf. Series **808** (2017) 012008 doi:10.1088/1742-6596/808/1/012008

Diagnosing pathology of chest organs at early stages is complicated due to overlapping views of the lung and bone tissues. By using different X-ray energy ranges tissues with different absorption factors are separated and their separate images are received. This technology ensures that pathology is revealed at the early stage during routine screening.

Using the exponential law of radiation attenuation, a system of two equations for low (LE) and high (HE) energy levels can be written. The Rayleigh and Compton scattering components are negligible.

$$I^{\mathrm{HE}} = I_0^{\mathrm{HE}} \exp(-\mu_{\mathrm{soft}}^{\mathrm{HE}} t_{\mathrm{soft}} - \mu_{\mathrm{bone}}^{\mathrm{HE}} t_{\mathrm{bone}}); \quad (1)$$

$$I^{\mathrm{LE}} = I_0^{\mathrm{LE}} \exp(-\mu_{\mathrm{soft}}^{\mathrm{LE}} t_{\mathrm{soft}} - \mu_{\mathrm{bone}}^{\mathrm{LE}} t_{\mathrm{bone}}), \quad (2)$$

where I и I_0 are intensity values after and before passing the object (corresponding to the image brightness in terms of digital radiography), μ is the attenuation factor, and t is the object thickness in centimeters.

Generally, the attenuation factor depends on the density of the electrons in the atom and the material atomic number.

$$\mu(\mathrm{E})_{\mathrm{P}} = \rho_e C_{\mathrm{P}} \frac{Z^3}{E^3}, \quad (3)$$

where ρ_e is the density of the electrons in the material atoms; E – energy; Z – material atomic number; $C_{\mathrm{P}} = 9.8 \cdot 10^{-24}$. The atomic number for the soft and bone tissues is 7 and 20 accordingly.

The anode voltage used in medical X-ray equipment varies from 40 to 150 kV. The following ranges are chosen for the double-energy technology: 50–70 kV and 120–140 kV. This provides for spacing the energy beams, maintaining the required quality of radiation, and extending the X-ray generator's useful life. Additional filtering is used to suppress low-energy components.

When solving a system of linear equations (1)–(3), the brightness values for soft and bone tissues can be represented as follows:

$$I_{\mathrm{soft}}^{\mathrm{DE}} = \frac{I^{\mathrm{HE}}}{I^{\mathrm{LE}} \omega_{\mathrm{s}}};$$

$$I_{\mathrm{bone}}^{\mathrm{DE}} = \frac{I^{\mathrm{LE}} \omega_{\mathrm{b}}}{I^{\mathrm{HE}}},$$

here ω_s and ω_b are suppression factors for bone and soft tissues, equal to the ratio between the tissue attenuation factors for high and low energies. The values of the factors depend on the study environment, filtering, exposure parameters and the patient's anatomic features. The following values were obtained experimentally: $\omega_s = 0.25 \pm 0.10$ and $\omega_b = 0.60 \pm 0.10$ [3].

Below, sample thorax images of a Kyoto Kagaku PBU50 phantom received using this technology by a Telekord-MT-Plus device made by CJSC Medical Technologies Ltd are presented (figure 3). Additional radiation filtering was used (1 mm Al and 0.2 mm Cu), with the grid (R12, 80 lp/cm) and the factors of 0.25 and 0.70.

Two types of parameters affect the final image quality: exposure and reconstruction parameters. Exposure parameters include: presence of the grid; type and value of additional radiation beam filtering; ratio between the radiation doses of both exposures. Reconstruction parameters include: suppression factors for bone and soft tissues; noise attenuation method used in the reconstructed image; noise attenuation factors.

Using modern tomosynthesis and double-energy technologies increases the information capacity of the study significantly and improves diagnostic capabilities at early stages. The main advantage of these technologies is the ability to use them with remote-controlled gantry tables with a flat-panel dynamic digital X-ray detector, without significant adaptation of the hardware, as well as with modern devices having two workstations.

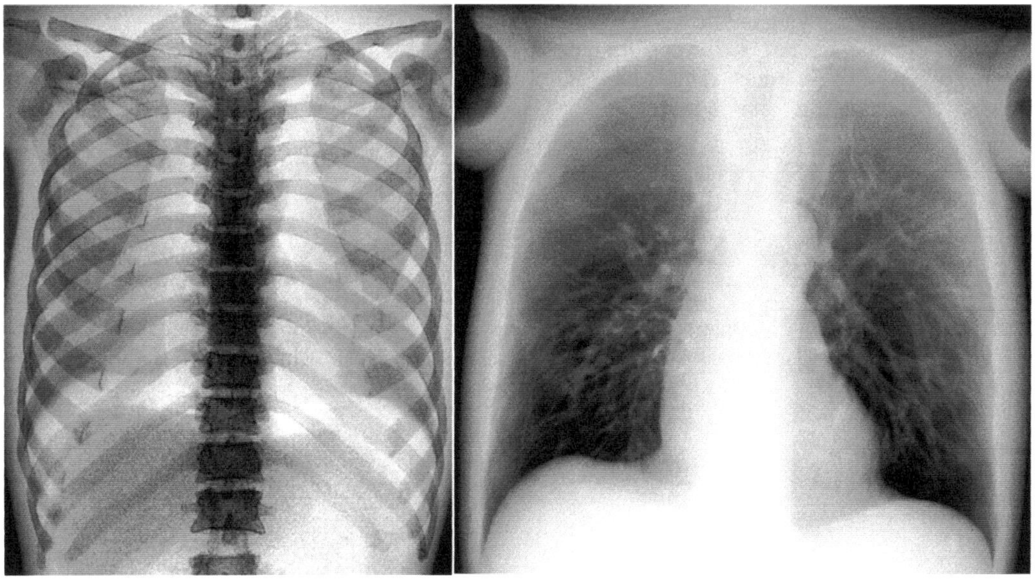

Figure 3. Sample reconstruction of bone (left) and soft (right) tissues.

It is assumed that, when introduced into medical practice, double-energy and tomosynthesis technologies will in some cases make additional CT examinations redundant. This will help relieve CT rooms of the burden of routine examinations and make them accessible to more patients.

References

[1] Sabol J M and Heckel B 2012 *Proceedings of conference of European Society of Thoracic Imaging (London)*

[2] Gomi T, Hirano H, Nakajima M and Umeda T 2011 *Journal of Biomedical Science and Engineering* **4** 443–53

[3] Richard S 2008 *Optimization of imaging performance and conspicuity in dual-energy x-ray radiography* (Toronto: University of Toronto)

IOP Conf. Series: Journal of Physics: Conf. Series **808** (2017) 012009 doi:10.1088/1742-6596/808/1/012009

Quality assessment of digital X-ray chest images using an anthropomorphic chest phantom

A V Vodovatov[1], I G Kamishanskaya[2], A A Drozdov[2] and C Bernhardsson[3]

[1] Saint Petersburg Research Institute of Radiation Hygiene n. a. Professor P V Ramzaev, 197101, Saint Petersburg, Russia
[2] Saint Petersburg State University, Faculty of Medicine, Saint Petersburg, Russia
[3] Medical Radiation Physics, Department of Translational Medicine, Malmö, Lund University, Skåne University Hospital, 205 02 Malmö, Sweden

E-mail: vodovatoff@gmail.com

Abstract. The current study is focused on determining the optimal tube voltage for the conventional X-ray digital chest screening examinations, using a visual grading analysis method. Chest images of an anthropomorphic phantom were acquired in posterior-anterior projection on four digital X-ray units with different detector types. X-ray images obtained with an anthropomorphic phantom were accepted by the radiologists as corresponding to a normal human anatomy, hence allowing using phantoms in image quality trials without limitations.

1. Introduction

Accurate measures of both clinical image quality and assessment of patient radiation risk are essential for successful optimization of medical imaging with ionizing radiation. Optimization in diagnostic radiology is equivalent of finding the image acquisition technique that maximizes the perceived information content and minimizes the radiation risk or keeps it at a reasonably low level [1]. All the changes made to the examination protocols should be applied only after it is confirmed that the revised exposure settings do not degrade diagnostic features of the image.

Currently, there are a variety of different techniques for the assessment/measurement of image quality [2–5]. Objective methods, based on the description of the physical characteristics of the image and their relation to the exposure, are usually performed with test-objects and simple physical phantoms. Subjective (or observer performance) methods, based on the visual perception of the whole image or parts of it, are usually performed with anthropomorphic phantoms for medical imaging or with real patient data [2, 5]. However, results obtained using the objective methods are more accurate and reproducible; they allow both measuring the primary physical characteristics of the imaging system or the overall system performance [3, 4]. Unfortunately, establishing a direct relation between those physical image characteristics and clinical image quality is difficult [3, 5]. On the other hand, the observer performance methods allow evaluating the quality of the whole imaging chain, thus measuring the image quality. Although there is a risk that these methods are influenced by a subjective nature of the observer, they consider the professional opinions and demands of radiologists [2, 5].

Observer performance methods are either based on the detection of the lesions (Receiver Operating Characteristics, ROC, and its variants) [4], or on the assessment of the visibility of the anatomical structures (visual grading analysis, VGA, or Image criteria, IC), the latter being more easily implemented in the clinical practice [5–7]. Current set of image criteria for conventional radiography of adult patients have been presented by the European Commission [8]. These quality criteria have however not been updated since they were presented in 1996 and were primarily designed for the analogue X-ray equipment. Russian national radiological guidelines [9] suggest performing chest imaging in frontal projection on 80–100 kV, and the majority of the Russian X-ray unit vendors comply with those standards. Using relatively low tube voltage compared to the European and International practices (120–140+ kV) leads to fewer possibilities of lesion detection and yields to a higher exposure of the patient. Additionally, low tube voltage requires relatively high tube current-time product, resulting in a significantly higher exposure of the patient.

Hence, the main aim of the current study was to evaluate the clinical quality of chest images acquired at different tube voltages, and to select the optimal one based on the radiologist's opinion. The second aim of the study was to select the proper image quality assessment methodology that would be flexible and could be used as a tool for different optimization procedures based on end users (radiologists) subjective image quality assessment.

2. Materials and methods

The initial image quality criteria's were taken from Tingberg et al. [2], Uffmann et al. [7] and the European Commission recommendations [8]. Additional criteria were added by radiologists from the participating hospitals with more than 10 year working experience with digital X-ray units. Twelve criteria were selected, corresponding to the different anatomical regions of the chest. Complex anatomical regions (e.g. visibility of the vascular pattern in the peripheral areas of the lungs superimposed on ribs) were divided into the sub-regions.

The use of an anthropomorphic phantom instead of real patients allowed investigating the X-ray tube parameters otherwise impossible. Hence, in the present study it was possible to perform investigations of a variety of different protocols using the same object. The study was conducted using the Multipurpose Chest Phantom N1 "Lungman" (Kyoto Kagaku Co., Japan) anthropomorphic phantom [10, 11], simulating the anatomy of the chest of a healthy male (174 cm height, 75 kg weight). No pathologies (e.g. nodules, lesions, etc.) were added. During the image acquisition, the phantom was positioned in a standard posterior-anterior (PA) position for a patient undergoing chest examination according to a typical clinical protocol [5, 9]. Four digital X-ray units with different detector types that are commonly used for chest screening examinations (all manufactured by CSJC "Electron", Russia) commonly used for chest screening examinations in the radiology department of the Mariinsky hospital were selected for the study. Chest images were acquired with an automated exposure control at 3 different tube voltages: standard for Russia (90 kV), lowest used in clinical practice (60 kV) and highest used in clinical practice (120 kV). As a result, four sets (one set for every X-ray unit) of three images were constructed. Dose-area product (DAP) was measured using DRK-1 DAP-meters (DOZA, Russia), integrated into the X-ray units. The DAP-meters were calibrated by the vendor prior to the study. Effective dose was calculated using PCXMC software (STUK, Finland), using tissue weighting factors from ICRP publication 103 [12].

The trial on diagnostic image quality assessment was conducted at the radiology departments of two major university hospitals in Saint Petersburg: State Mariinsky hospital and Military Medical Academy. Thirteen radiologists from both hospitals (ten from Mariinsky hospital and three from Military Medical Academy) were selected as experts for the trial: six of them had less than 10 years of experience and seven of them had more than 10 years of experience, composing a representative group for the Russian radiological practice in the region. All the participants had a prior experience of working on the selected models of X-ray units. The phantom images were exported as DICOM files from the hospital PACS server. For each X-ray unit three images acquired at 60 kV (minimal tube voltage allowing chest imaging), 90 kV (median for the tube voltage range) and 120 kV (maximum tube voltage allowed by the generator capabilities) were selected. The images were processed by the X-ray units' software according to the standard settings for the chest protocol. Manual post-processing of the images (i.e. windowing, filtration) was disallowed due to the methodology of the trial (segmentation and presentation of the resulting images). The regions and sub-regions of the images, corresponding to the selected criteria, were cut from the original images, shuffled and merged. This resulted in 25 separate triplets of images for each X-ray unit (figure 1). The resulting images were presented to the observers on a dedicated workstation with a diagnostic monitor (NEC MD21GS-2MP-BB, 21 inch, 3000×3000 pixel resolution), calibrated by the vendor.

Preliminary tests using the IC assessment method indicated that the digital images acquired on the X-ray units participating in the trial complied with the proposed criteria regardless of the parameters of image acquisition (tube voltage, specifically). All the images were marked as acceptable or above average quality. The current study was performed using a combination of the VGA and IC methods, thus allowing comparing the visibility of the suggested image criteria between the images acquired on different tube voltages. For each participating X-ray unit, images acquired on different tube voltages with the AEC were extracted from the PACS. After that, the areas corresponding to the relevant anatomic regions and sub-regions were cut from the original digital images. Further, for each region and sub-region cut, areas were shuffled to form a random sequence and merged together into one resulting image (triplet) for evaluation. An example of such resulting image is presented in figure 1.

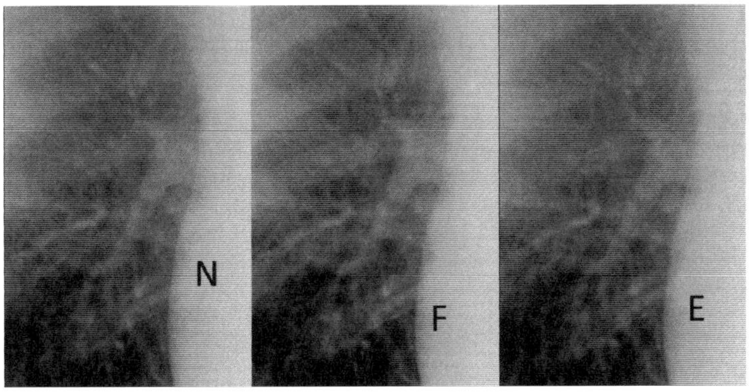

Figure 1. Example of the resulting image for the criterion 3 (Hilum of the lung) for the FC X-ray unit. Segments corresponding to the anatomical region of the hilum were cut from the chest images acquired on a different tube voltage (60, 90 and 120 kV), shuffled and merged. The letters (N, F, E) encode the tube voltage.

Instead of assigning a score based evaluation system on the quality of the criteria visualisation, the observers were required to rank the presented anatomic regions (figure 1), assigning first, second and third places. In case of a tie, each image was assigned the closest highest score. Each image segment received a score according to an assigned place. The image score was calculated as the sum of all the points for the segments. The image with the least overall score was considered the best.

A guidance on the selected image quality assessment method was distributed among the observers. It included a list of the criteria with the corresponding anatomic regions marked on the images of the chest. A series of test trials were performed to assure that the selected observers became familiar with the proposed methodology of the image quality assessment. The image rankings were converted into scores and processed using Kruskal-Wallis non-parametric test, using Statistica 10 software. The dependent variables included the rating scores for the image criteria; tube voltage was selected as the independent variable. Differences were considered significant with a $p < 0.05$.

3. Results and discussion

The overall image scores per X-ray unit were obtained by summing the scores for the separate image criterion for each observer. The results are presented in figure. 2 for each of the selected tube voltages (60, 90 and 120 kV) and for each X-ray unit. The difference between different tube voltage was insignificant only for the FC X-ray unit (Kruskal-Wallis test $p = 0.6$). The 60 kV protocol was selected as the best for the ARC1 X-ray unit and the 120 kV protocol was selected as the poorest for the KRT X-ray unit. No statistically significant differences between the remaining protocols were observed for those units. For the ARC2 X-ray unit (the only unit with an indirect flat-panel detector) statistically significant differences were observed between all three tube voltages. Also for this unit, the 60 kV image was selected as the best and the image acquired with 120 kV as the worst. Mean observer scores for the separate image criteria (with the image scores for the sub-regions averaged into the scores for the whole region) and the data on the significance of the differences are presented in figure 2. The effective dose was calculated based on the DAP for the reference adult [12].

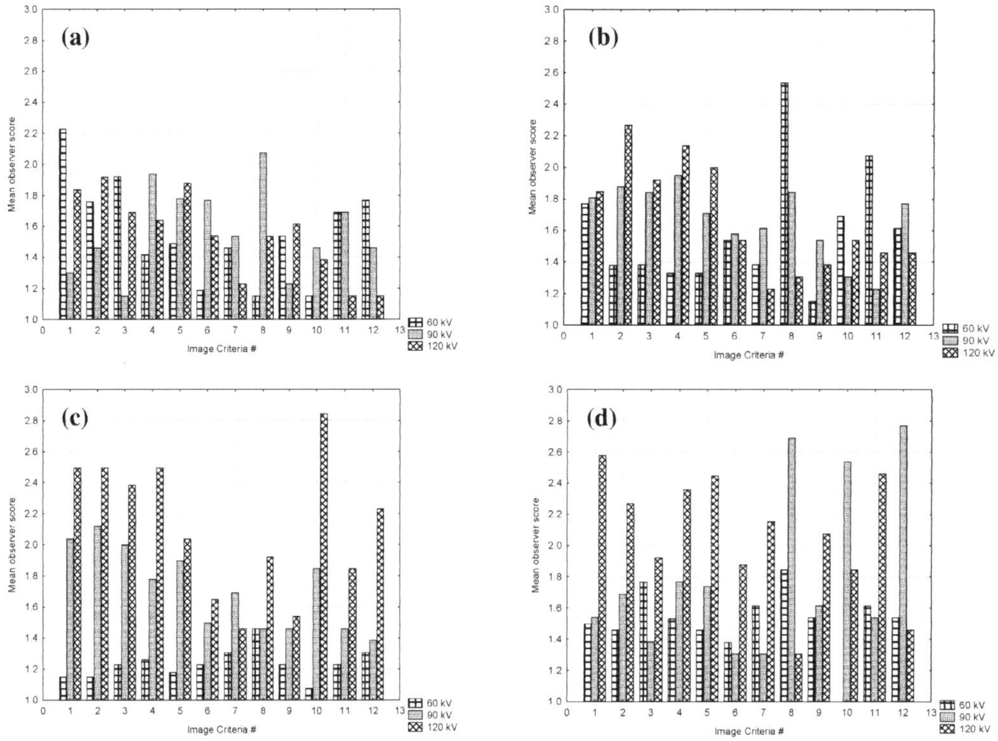

Figure 2. Mean observer scores for the selected image quality criteria for the FC (**a**), ARC1 (**b**), ARC2 (**c**) and KRT (**d**) X-ray units, respectively. Image scores for the sub-regions were averaged into the scores for the whole region. The best image quality corresponds to the lowest image score.

The Russian radiological practice lacks systematic approach to the image quality description and assessment, as the majority of the radiologists tend to evaluate the whole image basing their decision on the initial visual impression. Another factor is that most radiologists and operators work simultaneously with both analogue and digital X-ray units. Unsatisfactory digital images are considered to be acceptable because they still provide better diagnostic quality compared to images acquired on analogue units. Hence, the general demands on digital image quality are lower. In this case, using only the Image Criteria method would lead to the misinterpretation of the results. As mentioned before, in the preliminary trials based on the IC method, all images were marked as acceptable or above average, with no statistically significant differences between the combinations of the parameters of the examination (data not provided). Allowing the observers to rank the presented images by the visibility of the different anatomical structures (relating them to the different image quality criteria) tend to be the most feasible choice [5, 13].

Segmentation of the images, presenting only the cropped parts, corresponding to a specific anatomic region is a forced measure needed to perform the present study. Although it complicates the preparation of the trial, and disallows manual post-processing of the images, it is currently the only solution to force the observers to focus on the specific regions of the image, thus making the result reliable [2]. The use of the anthropomorphic phantoms in the trials was accepted by the radiologists, as long as the image quality assessment was focused on the visibility of the normal anatomy. Using the phantoms allowed also comparing different imaging protocols and X-ray units without any limitations due to the ethical considerations [10]. Nevertheless, the set of the chest image quality criteria used in the present study was originally developed for real patients.

It is generally accepted, that the diagnostic image quality decreases with the increase of the mean energy of the X-ray beam due to the significant reduction of image contrast [5, 7]. The results of the current trial confirm this. For three out of four X-ray units and for the majority of the observers, 60 kV and 90 kV images had the lowest, and 120 kV images – the highest mean image scores (figure 1). For the FC X-ray unit the difference between three imaging protocols were not statistically significant. It is worth mentioning that the clear differences between the protocols, hence indicating the consensus of the observers, were visible only for the ARC2 X-ray unit with a flat-panel detector. The 60 kV and 90 kV or 120 kV protocols were similar for the X-ray units using a CCD-matrix detector.

It was not possible to determine the statistically significant differences for the majority of the separate image criteria, although they follow the same trend. It is visible from figure 2 that 60 kV images were mainly marked as best and 120 kV images – as worst, for the majority of the image criteria. According to our expectations the image criteria's were subdivided into three groups: best visible on low tube voltage; best visible on high tube voltage and criteria not affected by tube voltage at all. That grouping was based on the anatomical structure and the location of the relevant area. The anatomical structures located behind large and dense objects (i.e. thoracic aorta, located behind the sternum and the heart) were expected to be better visualized using high tube voltage. On the contrary, subtle and soft structures (pulmonary parenchyma) were expected to be better visualized using low tube voltage. It was expected, that the observers scores would be influenced by the differences in the visualization. According to the results (figure 2) low (60 kV and 90 kV) tube voltages were the best for all the three groups, contradictory to the results expected for group 2 and 3. Only the visualization of criteria 7 and 9 conformed to our expectations. It was not possible to identify any significant differences between the radiologists from both of the participating hospitals, or with different working experience. The diversity of the opinions of the observers can be explained by the personal preferences due to the different levels of education and training.

The preference of the low tube voltage for the majority of the criteria can be explained by different factors. Firstly, radiologists from the both hospitals commonly use 70–100 kV tube voltage range for the real patients, thus they intentionally or unintentionally preferred segments of images with similar appearance. Secondly, the radiologists interpreted a "good/best image" not as an "image with maximum information content", or, in case of the current study, as an "image with the best representation of the normal anatomy of the chest" but as an "image with best appearance". Despite the efforts to focus the observer's attention on the individual anatomic areas and particular anatomic details, it can be concluded that the majority of them still tend to evaluate the whole image presented. The decisions were made without considering the actual visibility of certain anatomical structures. This trend can be changed by additional training of the radiologists, introducing the concepts of image quality and quality criteria with an addition of the algorithmic approach to the image evaluation [13]. The visual appearance of the presented image segments acquired on 60 kV was more detailed in contrast compared with the others. That can be explained by the energy of the beam corresponding to the optimal energy for the CCD and CsI based flat panel detectors, complemented by the unsatisfactory X-ray unit image processing. The results for the ARC2 X-ray unit correspond to the results described by Uffman et al. [7], with low tube voltage (90 kV) protocol being significantly better compared to 121 and 150 kV protocols for the flat-panel detector system.

In order to fully evaluate three different imaging protocols, the patient dose should be considered as well. DAP and effective dose for the 60 kV imaging protocol are higher by a factor of 1.5–2 and 2–2.5 compared to the 90 and 120 kV, protocols respectively. In this case, using high tube voltage is beneficial, considering that screening examinations are performed mostly on healthy patients. The most viable approach to manage patient dose is to use the low-dose (high voltage) protocols for the X-ray units with the CCD-matrix detector and try to lower the patient dose for the low tube voltage protocol by adjusting the AEC settings and upgrading the image processing software for the X-ray unit with a flat-panel detector. Another solution would be to adjust the exposure of the detector at the selected values of tube voltages, keeping the effective dose to the patient on the same level, and perform an additional image quality assessment study. That would allow for the assessment of the tube voltage, without considering the dose of the patient.

4. Conclusions

The results of our trial on clinical image quality indicate the following. We propose a diagnostic image quality assessment method using an anthropomorphic phantom. This method is based on the combination of the image criteria and visual grading analysis. It allows to evaluate the quality of the whole image or the separate anatomic areas, and can be used for the comparison of the different imaging protocols between the different X-ray units. The comparison of the chest images of the anthropomorphic phantom indicates that images, acquired on low tube voltage (60 and 90 kV) were preferable compared to images acquired on high tube voltage (120 kV);

High variations in the scores indicates a significant difference in education and training of the observers, both in requirements to the diagnostic image quality and experience in similar trials. Using the imaging protocols preferred by the majority of the radiologists with the current exposure settings will lead to significant additional exposure of the patient (up to the factor of 2–2.5 compared to high tube voltage protocols). We propose to to use the low-dose (high voltage) protocols for the X-ray units with CCD-matrix considering the additional exposure of the patient and high variation in the image quality assessment. Low tube voltage should be used with X-ray units utilizing flat-panel detectors after the attempts to lower the patient dose by adjusting the AEC settings and upgrading the image processing software.

Acknowledgments

The authors would like to thank the staff of the radiology departments of the Saint Petersburg State Mariinsky hospital and Saint Petersburg Military Medical Academy for the possibility to use the X-ray equipment and the participation in the study.

This publication is part of a research work at Lund University, according to Swedish Institute scholarship (00110/2015) for A V Vodovatov. This work is partly financially supported by the Swedish Radiation Safety Authority (SSM).

References

[1] Wouter J H V, Lucia J M K and Jacob G 2009 *Eur. Radiol.* **72** 209–17

[2] Tingberg A 2000 *Quantifying the Quality of Medical X-Ray Images: An Evaluation Based on Normal Anatomy for Lumbar Spine and Chest Radiography* (Malmö: Lund University)

[3] Tapiovaara M 2005 *Rad. Prot. Dosim.* **1–3** 116–9

[4] Chakraborty D P and Berbaum K S 2004 *Med. Phys.* **8** 2313-30

[5] Image Quality in Chest Radiography 2003 *Journal of the ICRU* **3**

[6] Lanhede B, Båth M, Kheddache S et al. 2002 *Br J Radiol* **75** 38–49

[7] Uffmann M, Neitzel U, Prokop M, Kabalan N, Weber M, Herold C J, Schaefer-Prokop C 2005 *Radiology* **235** 642–50

[8] Carmichael J H E 1996 *European Guidelines on Quality Criteria for Diagnostic Radiographic Images* (Luxembourg: EUR-OP)

[9] Abduraimov A B 2013 *National guidelines on X-ray diagnostics and therapy* (Moscow: GEOTAR-Media)

[10] DeWerd L A and Kissick M 2014 *The Phantoms of Medical and Health Physics. Biological and Medical Physics, Biomedical Engineering* (New York: Springer New York)

[11] Chest phantom N1 user manual: http://www.kyotokagaku.com/products/detail03/pdf/ph-1_manual.pdf

[12] Tapiovaara M and Siiskonen T 2008 *PCXMC. A Monte Carlo program for calculating patient doses in medical x-ray examinations* (Helsinki: STUK-A231)

[13] Folio L R 2012 *Chest imaging. An algorithmic approach to learning* (New York: Springer)

IOP Conf. Series: Journal of Physics: Conf. Series **808** (2017) 012010 doi:10.1088/1742-6596/808/1/012010

Current status and prospects of X-ray microscopy

V D Gelever, E Yu Usachev and A A Manushkin

Moscow Technological University (MIREA), 119454, Moscow, Russia

E-mail: gelgan@yandex.ru

Abstract. Design and main constructive features of a new combined device for X-ray electron and X-ray microscopy for nanometer diagnostics are considered in relation to the current status of modern X-ray and electron microscopy.

1. Introduction

Currently, X-ray microscopy enables to conduct non-invasive research on the micron level of the different materials and devices in various fields of technology. With increasing resolution of X-ray machines, they are increasingly being used for scientific research at the nanoscale in X-ray microscopes.

In the X-ray machines, sealed and dismountable X-ray tubes are used. The formers have a low cost, small size, but a limited life time and may provide low resolution, worse than 10 μm. This is due to the fact that all of the current from the cathode is focused on the target. Even the use of scaling down magnetic lenses placed outside the tube does not allow exceeding this limit due to lens aberrations without the appropriate aperture on the tube axis.

In the dismountable tubes for focusing the electron beam after cathode, 2–3 electromagnetic lenses were applied, in which the relevant restrictive and aperture diaphragms are established to optimize the aberration on the axis. In principle, such an electron-optical systems are used in electron microscopes to obtain nanoscale electron beams. In the case of X-ray microscopes, X-ray source is the generation region (focal spot) formed in the target which is being broadening due to the path length of the electrons. Further, X-rays output rate from the target is lower by several orders of magnitude, than the corresponding rates for the secondary electrons. Therefore, X-ray imaging requires high current electron beams, which corresponds to their large size and low resolution.

A high resolution (50 nm) is achieved in laboratory nanoXCT-100 systems (X-Radia, USA) with a rotating anode tube and X-ray optical elements (figure 1). Higher resolution (20–30 nm) on biological objects was obtained at synchrotrons using X-ray optics [1]. A number of foreign companies produce high-voltage microfocus inspection apparatus with collapsible X-ray tubes having a limiting resolution at 0,1–0,5 μm, that allow research and inspection on the micron level. The majority of them are performed in a single module embodiment, when almost all the constituent elements are in a single protective housing. Therefore, they have a large overall size to the size in three dimensions 1–2 m with volumes 6–8 m^3 at 2–3 ton weight. These devices are mainly designed to control large circuit boards. In general, due to their design and dimensions, it is difficult to call such devices as microscopes and they are rarely used in scientific research.

Among these devices outstands a bit XD-7600NT – Dage (England) with the limiting resolution 0,1 μm (figure 2). X-ray tube used in it, is advertised as a revolutionary step in the development of the tubes, since it combines the advantages of conventional sealed and dismountable tubes. The tube is called sealed-transmissive. In it, a small magnetic drive and a gettering pumps are connected to a system of magnetic focusing lenses. There is also an indoor high vacuum gate through, which the primary backing pumping, is conducted by the manufacturer and launch o high vacuum pumps. In such a tube, high vacuum is maintained permanently and only if necessary, the running of magnetic drive pump occur. The gain in production costs compared to conventional dismountable tubes is minimal, since only the backing pump is excluded from the cost. The pumping time is little only in the case of small intervals between switching on and off. And at large period there is a possibility not to obtain a high vacuum, due to leakage into a vacuum volume. In practice, in some cases it occurs. As a result, it is necessary to send the tube to the manufacturing company to produce a vacuum. In addition, the cost of such a tube is still much higher, than the cost of a sealed tube.

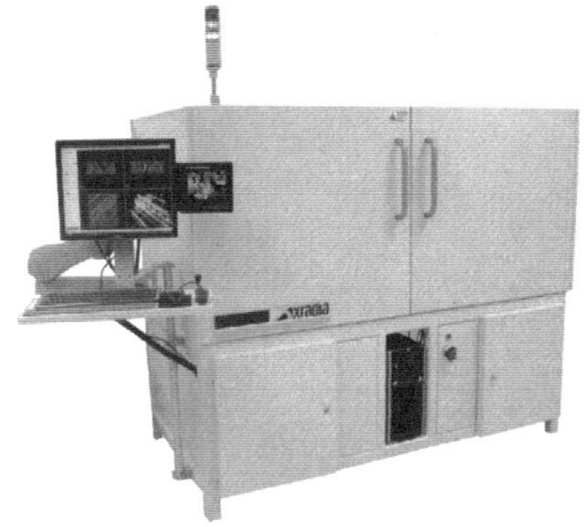

Figure 1. Laboratory nanoXCT-100.

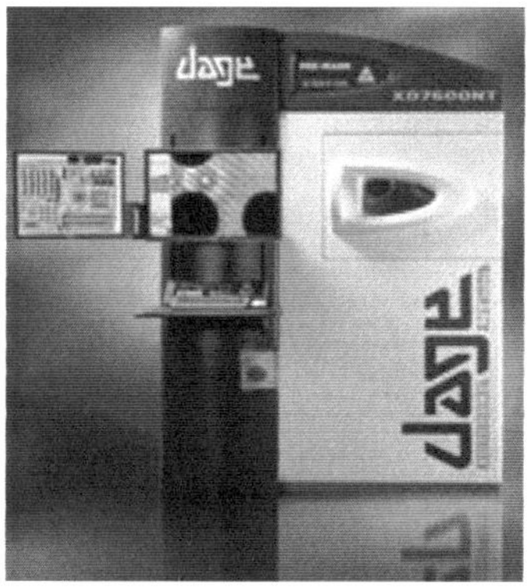

Figure 2. XD-7600NT – Dage (England).

At present, in the Russian Federation, X-ray machines at the dismountable tubes are not made serially, but there is some experience in this area. For example, until the 90s the Vyborg instrument-making plant manufactured X-ray microscopes (MTP-7 and others) with a resolution of 5 μm. It should be noted, that with modern X-ray detectors and transmission targets, a resolution less than 1 mm would be obtainable at them. In addition, at the Moscow technology university MIREA, the X-ray measuring microscope (XMM) was made to control the thickness of layers in carbon like nuclear fuel elements, working at accelerating range 5–30 kV with 0,2–0,3 μm resolution (figure 3).

In principle, it is expedient to manufacture microfocus devices in the modular embodiment as MTP-7, where in in the protective chamber there are only objects moving mechanisms and detectors. Dismountable tube and power supply units are out of the camera. Herewith, the devices will have smaller size and cost. Figure 4 shows an experimental model using a camera on the MTP-7.

Figure 3. XMM.

Figure 4. Shielding camera (X-ray dismountable tube – at left, object table – at back, detectors – up or at right).

2. Hybrid nanoscope

For X-rays nanometer resolution at the level of tens of nanometers, it has been developed a hybrid nanoscope (HN), which optimally combines scanning electron microscope (SEM) and X-ray microscope (XM). To maximize the resolution parameters in both modes, an optimized system of magnetic lenses with interchangeable pole pieces was used to focus the electrons [2]. In the SEM mode, an object is set under the electron beam, and with secondary electron detector (SED) before the last (objective) lens in a scanning mode the secondary electrons from the object are recorded. In XM mode, a target on a vacuum-tight substrate is set under the electron beam. For accurate and fast focusing of electrons on the surface of the target, over the entire operating range of resolutions an image of the target surface is obtained in secondary electrons using a SED, while scanning electron at a target [3]. XM can work in projection and scanning (reverse) mode (figure 5).

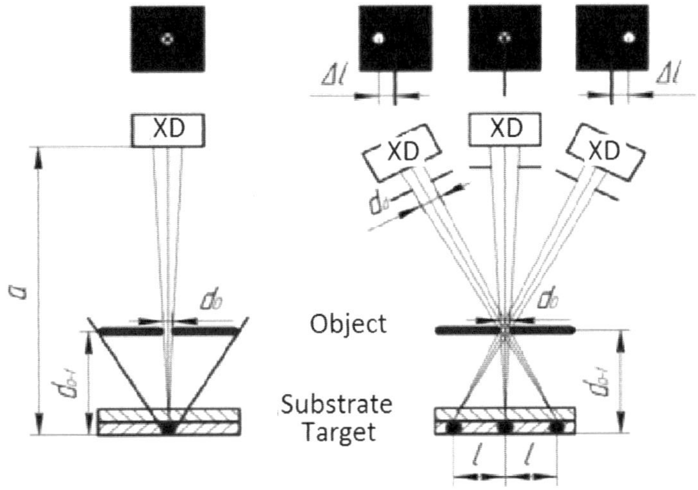

Figure 5. Projective and scanning (with 3 detectors) modes (d_o – diameter of hole in the object); $d_{o\text{-}f}$ – object-focus distance; a – target-detector distance; XD – X-ray detector; d_d – hole diameter at the detector input).

To obtain high adsorption contrast for the nanoscale details of objects, it is necessary to use the accelerating voltage on 5–15 kV level. In this case, the resolution level of 20–30 nm can be achieved in near focus mode, when a target is laid on a sub-micron and micron substrate of Be, Si, Si_3N_4, C, etc. This allows placing objects on micron and submicron distances from the focal spot. In the microfocus X-ray microscopy, there are commonly used the substrates with thicknesses at the level of hundreds of microns. While using thinner substrates, the density of X-ray flux rises at both the object and the detector. This makes it possible to compensate for a decrease in the intensity of X-ray source at the nanoscale focal spot. In addition, for small distances between the subject and the focus, also a phase contrast mode can be effectively realized. The material and the thickness of the target influences greatly on the intensity of the soft X-rays flowing after the substrate. The use of materials with high density, in which electrons have a smaller path length, reduces beam broadening and increases the radiation intensity. If the path of the electrons will be more than the target thickness, a part of the electrons pass through the target without producing X-rays, while this slightly reduced broadening of the electron probe. With the thick targets it may occur that the part of the target, to which the electrons are not reached, will substantially absorb X-rays from the focal spot. Therefore, it is necessary to optimize the thickness of the target in accordance with the accelerating voltage. The resolution in the X-ray is also significantly affected by the thickness and chemical composition of objects. As a result, if you choose the optimal actual test object, the parameters of the target and the accelerating voltage, it is possible to obtain high resolution at the level of tens of nanometers.

IOP Conf. Series: Journal of Physics: Conf. Series **808** (2017) 012010 doi:10.1088/1742-6596/808/1/012010

3. Results and discussion

The five experimental samples (figure 6) were produced during the development of HN. Good initial results have been obtained in their test. Due to their technical parameters HN's allow to combine the key features and capabilities of the majority of imported electronic and X-ray microscopes. It is not easy an imports substitution, rather it is an import superior nanodiagnostics device. The advantage and performance of the HN can be seen, when comparing the images of objects with approximately the same size (figures 7, 8) made on different devices. And we must remember that the first two shots were taken at the upper level of the capacity. The pictures on the HN are not made under optimal conditions. The projection shooting mode was carried out on X-ray film and the scanning one – by a single detector. Going to X-matrix and the use of 10–20 detectors in raster mode will significantly increase resolution and image performance. It should be noted, that you can significantly increase the resolution due to the use in dismountable tubes of the field emission cathodes, as an electron source.

Figure 6. External view of electron-probe modules of HN (0–40 kV) (left) and low-voltage (0–15 kV) HN (LVHN) (right).

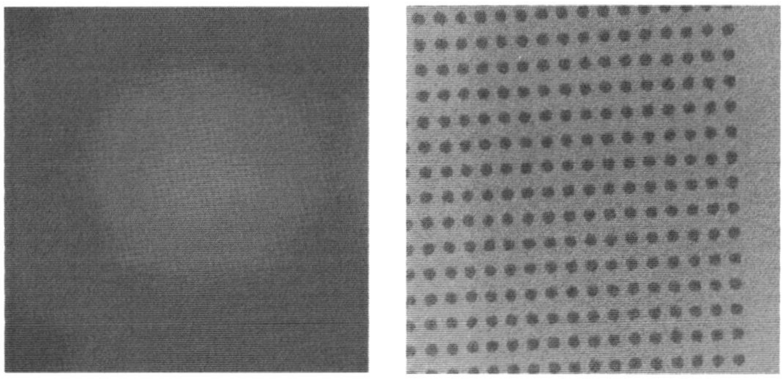

Figure 7. Metal balls with 20 μm diameter: sealed X-ray tube (left) and XD-7600NT (right).

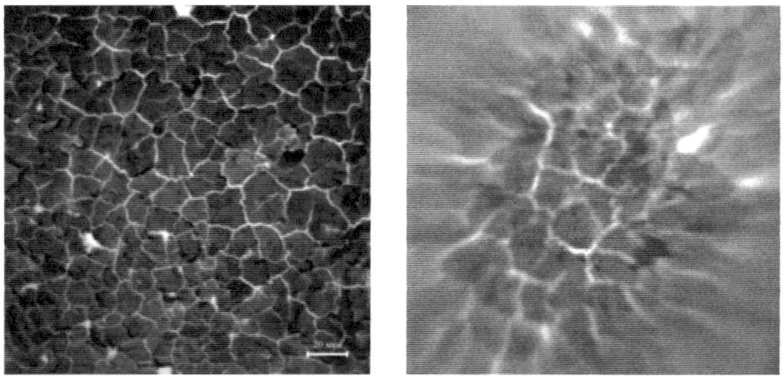

Figure 8. Re-film (grain size: 20–30 μm, gaps between them: 1 μm, film thickness: 2–3 μm): HN-projection mode (left) and HN-scanning mode (right).

4. Conclusion

It should be noted that the improvement of X-ray microscopes goes both in the direction of increasing of X-ray sources performance, parameters, as well as of increasing the parameters of detectors and X-ray optical elements. As a result, one can expect to go in the near future to the level of 10 nm.

References

[1] Chen Y-T et al. 2008 *Nanotechnology* **19** 395302
[2] Gelever V D 2000 *Bulletin of the Russian Academy of Sciences Physics* **8** 1584
[3] Gelever V D 2012 Patent RU 2452052
[4] Gelever V D, Usachev E Yu and Manushkin A A 2015 *Analitics* **3**

X-ray tubes for projection X-ray radiography of new generation

A A Podimsky[1] and N N Potrakhov[2]

[1] JSC «Svetlana-X-Ray», 198095, Saint Petersburg, Russia
[2] Department of electronic instruments and devices, Saint Petersburg Electrotechnical University "LETI", 197376, Saint Petersburg, Russia

E-mail: kzhamova@gmail.com

Abstract. This article presents the results of research in the field of creation of basic elements of high-voltage components for projection radiography. Design and characteristics of the three types of microfocus X-ray tubes of the new generation: with rotating and fixed anodes, and with external target are described.

1. Introduction

In the second half of last century for radiography of various objects in our country has become a widely used method of projection radiography, particularly in medical diagnostics [1, 2]. Clinical practice of those years has shown that for successful implementation of this method, the size of the focal spot of X-ray tubes should not exceed 0.1 mm or 100 μm. In accordance with GOST 22091.9-86 such X-ray tubes are classified as microfocus and in the future this method of projection radiography become called "microfocus X-ray radiography".

In the late '80s, thanks to the efforts of several schools of radiology, microfocus X-ray radiography emerged as an independent branch of medical diagnostics. And, one of the main roles was played by the Leningrad association of electronic instrument "Svetlana". Leading specialists of the individual design bureau of X-ray equipment "Svetlana" developed in association with radiologists of several large medical institutions in our country and successfully introduced into clinical practice microfocus X-ray machines of the family "Electronica", which until the end of the last century did not have foreign analogs.

2. Results and discussion

The researches, which were conducted at the department of electronic instruments and devices (Saint Petersburg Electrotechnical University "LETI") in collaboration with a number of leading national experts in the field of X-ray radiography, showed that the use of microfocus X-ray tubes in the formation of the X-ray image appears a number of features (effects). Main of them are: the effects of increasing the sharpness depth and contrast, the effects of pseudo 3D-image and phase contrast, the effect of reducing the exposure dose. However, due to the limit of power supplied to the target of microfocus X-ray tube by the electron beam of small cross section, the intensity of the generated radiation, in comparison with the "standard" X-ray tube, is small. This significantly limits field of application of microfocus X-ray radiography in such socially important areas of medical diagnostics like angiography, mammography, fluoroscopy, tomography, etc.

As it is known intensity *I*, generated by the X-ray tube, is connected with the parameters of X-ray tube by the following expression:

$$I = kizU^2, \qquad (1)$$

where I – is intensity of radiation, generated by the X-ray tube; k – coefficient of proportionality; z – atomic number of the material of the anode target; i – current of the tube; U – voltage on the tube.

In accordance with equation (1), the radiation intensity can be increased either by increasing a tube current, or by the voltage on it. Therefore, during the creation of new generation of microfocus X-ray tubes have been used both directions.

As it is known, on the tubes with a rotating anode current may be several orders of magnitude higher than in tubes with a fixed (stationary) anode [3]. Therefore, during the development of "high current" microfocus tube as base was chosen tube 15-40BD46-150, which is a kind of standard tubes in a class with a rotating anode. Modernization undergone two nodes of basic design – cathode and anode. As a result, in the cathode assembly has only one cathode, and to enable adjustment of the focal spot size the both outputs of cathode was electrically insulated from the focusing electrode. The composition of the modernized assembly includes: the cathode holder 1, directly heated cathode and the focusing electrode. The cathode 3 is made as a cylindrical spiral, two outputs of which are mounted in the holder, and both of outputs has ceramic insulators. Focusing electrode fixed on the cathode holder 1 (figure 1).

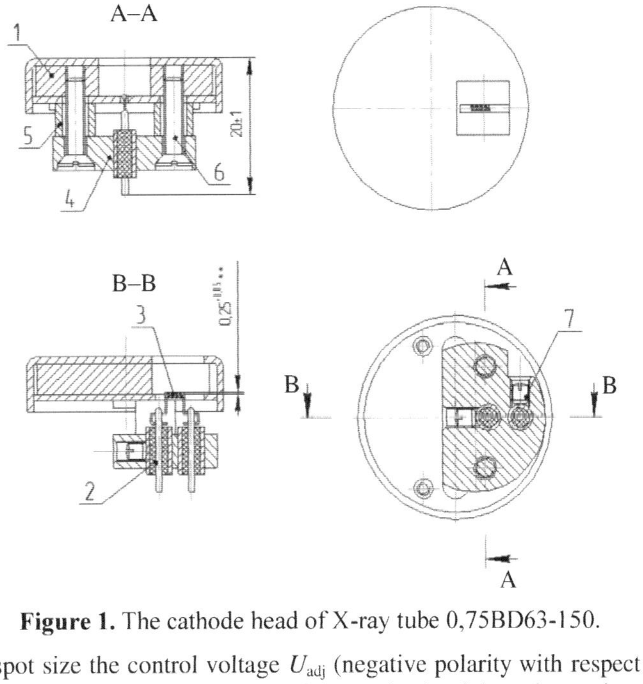

Figure 1. The cathode head of X-ray tube 0,75BD63-150.

To adjust the focal spot size the control voltage U_{adj} (negative polarity with respect to the cathode) is supplied to the focusing electrode. Depending on the magnitude of the voltage, the effective focal spot size will vary from a maximum 0.6×0.6 mm with a control voltage equal to zero to the minimum 0.1×0.1 mm with a control voltage to 100 V.

When upgrading anode node were tightened requirements to the value of bounce at node in the longitudinal and lateral (radial bounce) directions, and also to the magnitude of vibration and noise. Studies have shown that one of the causes of these effects – the lack of rigidity of the bearing housing and of the axis of the assembly rotation in the specified directions. In this basic design of the tube the radial clearance is comparable with the desired size of the focal spot. As a result of analysis and numerical estimates of allowable rotation speed of the anode assembly was chosen unseparatored ball-bearing assembly with a temperature-compensating gap type ZYS99.

The construction of the first domestic X-ray tube with a rotating anode with smooth adjustment of effective focal spot size 0.7BD64-150 shown in figure 2. Main characteristics of the tube:

– maximum voltage of 150 kV;
– maximum current of 12 mA;
– size of the focal spot at a bias voltage of 0 V – 0.6×0.6 mm;
– size of the focal spot at a bias voltage of 55 V – 0.1×0.1 mm;
– maximum rotation rate of the anode 9000 Hz.

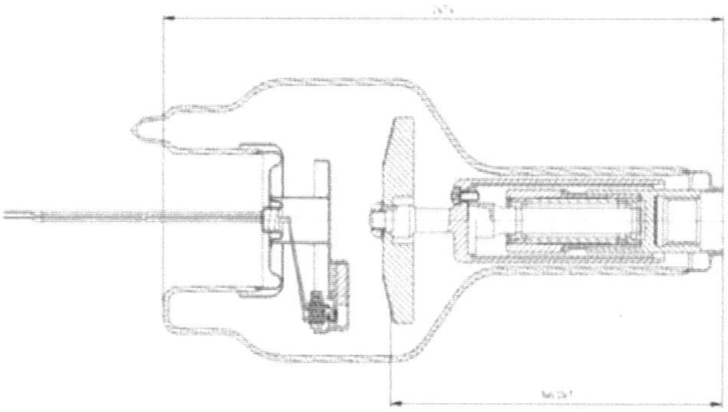

Figure 2. Microfocus X-ray tube with a rotating anode 0.7BD63-150.

For the development of microfocus X-ray tube at a higher voltage with a massive target located inside the vacuum container, was chosen tube design 1.2BPK21-200. Modernization has undergone the cathode assembly of basic design. Firstly, in order to provide the micron size of focal spot, instead of the cathode filament in the form a cylindrical spiral was used filament in the form of V-shaped loop. Secondly, in order to enhance the electron beam focusing capability by independent adjustment of the filament current and voltage on the focusing electrode, a second output of the cathode is also insulate from the focusing electrode. For this, both the cathode outputs is mount in the holder by screws, as in the tube with a rotating anode on ceramic insulators. The construction of the first domestic microfocus X-ray tube with a massive internal anode voltage of 200 kV 0.2BPM64-200 is shown in figure 3.

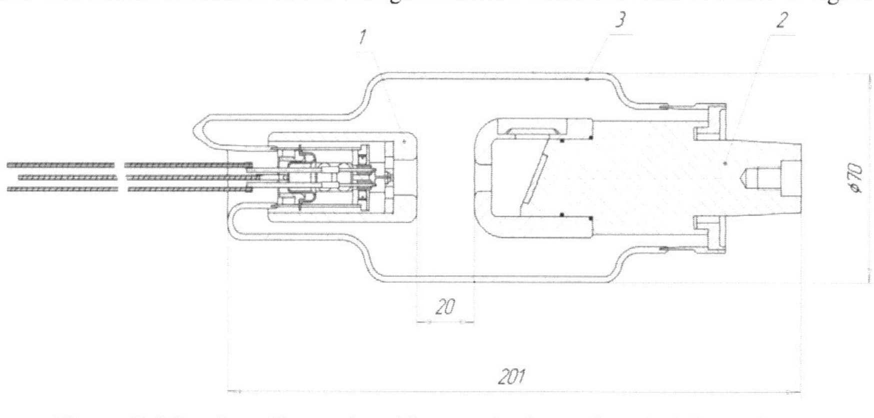

Figure 3. Microfocus X-ray tube with a massive internal anode 0.2BPM64-200.

The main characteristics of the tube:
– maximum voltage of 200 kV;
– maximum current of 1 mA;
– minimum size of the effective focal spot of 0.1 mm;

– dimensions: length 210 mm, diameter 70 mm.

For the development of microfocus X-ray tube at a higher voltage with the external target from the vacuum cylinder as a base was chosen tube BS16(I) with maximum operating voltage of 135 kV. During modernization of the base construction in order to increase the operating voltage to 150 kV by minimizing the electric field intensity in the high voltage gap of the tube were refined dimensions of the outside diameter of the focusing electrode of the cathode node and inner diameter of the anode screen. With the same purpose in the assembly of the tube was introduced the operation autofrettaging abrasive medium. Furthermore, in order to increase heat removal from external target and increase its loading capacity the anode material of the tube instead of stainless steel is replaced by copper. The design of microfocus X-ray tube with the external target from the vacuum cylinder for voltage 150 kV BS16(III) shown in figure 4.

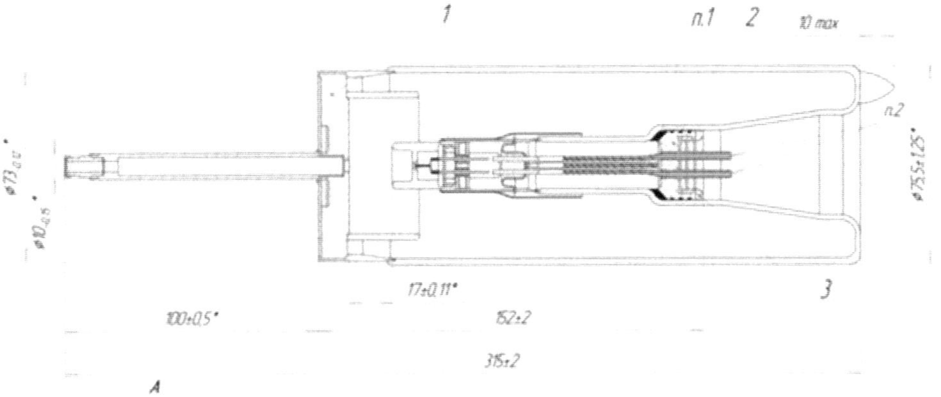

Figure 4. Microfocus X-ray tube with the external target BS16(III).

The main characteristics of the tube:
– maximum voltage of 150 kV;
– maximum current of 150 μA;
– minimum size of the real focal spot – 20 μm;
– dimensions of the balloon: length – 252 mm, diameter – 75 mm;
– dimensions of the external anode: length – 100 mm, diameter – 12 mm.

3. Conclusions

All three tubes of the new generation are designed for operation as part of a specially developed X-ray sources and devices: 0.7BD63-150 in the source IRD46 intended for medical diagnosis, 0.2BPM64-200 in a monoblock unit RAP-200M, BS16(III) in a monoblock unit RAP-150M.1 intended, in particular, for microfocus X-ray CT scanners family MRKT-01.

According to the preliminary assessment the implementation of these technical means will allow essentially extend the scope of Microfocus X-ray radiography in socially important fields of medical diagnostics, in the monitoring of the products and electronic components of the micro- and nanotech in all sorts of research.

References

[1] Blinov N N, Vasilyev A Y, Bessonov V B, Gryaznov A Y, Zhamova K K, Potrakhov E N and Potrakhov N N 2014 *Biomedical Engineering* **2** 58–61
[2] Potrakhov N N 2015 *Biomedical Engineering* **5** 237–40
[3] Podymsky A A and Potrakhov N N 2014 *Biomedical Engineering* **2** 78–80

Modernization of radiation detectors thickness gauge

I B Artemyev[1], B V Artemiev[2], Yu L Vladimirov[2] and L V Vladimirov[2]

[1] Bauman Moscow state technical university, Department IU4, ul. Baumanskaya 2-ya, 5, 105005, Moscow, Russia
[2] JSC Scientific-Research Institute of Introscopy "Spectrum", ul. Usacheva, 35, bl. 1, 119048, Moscow, Russia

E-mail: boris@artemiev.su

Abstract. Currently, there is a tendency in the industry by refusing isotopic radiation sources in favor of the X-ray machines. This is due to several factors, main among them radiation safety and maintenance problems, movement and disposal of gamma-ray sources. Compared to the gamma ray-source these devices have a number of disadvantages. The spectral energy distribution and therefore change in the spectrum as the radiation passes through the controlled material. Instability of radiation compared with gamma sources. All this complicates the use of X-ray sources for the materials thickness measurement with different chemical compositions.

1. Introduction

In metallurgical factories in Russia to address the problem of measuring the thickness of the rolled metal produced by widely used non-contact X-ray and isotope or radiation thickness gauges. The widespread use of these devices is justified their high metrological characteristics, performance and the lack of contact transducer with a moving surface in the metal rolling. Radiation thickness measurement – radiometric method for non-destructive testing for measuring thickness or surface density of the material and is based on the measurement of the parameters of ionizing radiation resulting from the interaction of the primary ionizing radiation controlled material according GOST 24034-80 "Non-destructive testing radiation" [1]. Park radiation thickness gauges requires constant maintenance, repairs and modernization. Currently, the problem of maintenance of the operated equipment in steel mills added a new task set by the leadership of our country – import substitution. To solve these problems at the same time the company "VIZ-Stal" Ekaterinburg decided to make work on the modernization of import FMM thickness gauges manufactured by "Messelektronik" German Democratic Republic (development of the '80s).

2. Formulation of the problem

The thickness meter is used the absolute measuring method (figure 1). Measuring clearance – 160 mm. Measured thickness for the steel from 0.1 to 2.0 mm. Random error (RE) measurements on a fixed sample thickness of 0.3 mm was 0.1 % compared to the measured thickness of hire. The equipment in the radiation source used isotope $Americium^{241}$, with the following main parameters:

- energy gamma rays of 60 keV;
- photon flux density at a distance of 1 m from the center of working surface $(7.0 \pm 2.1)\cdot 10^4$ $s^{-1}cm^{-2}$;
- Am^{241} maximum activity at the source 2.24 Ku ($8.3\cdot 10^{10}$ Bq);
- half-life 432.6 years;

– energy γ-quanta $E = 59.5$ keV (35.8%).

Sources "IGIA-5 M-1" produced by the Russian "Isotope" and can be purchased on the territory of Russia in the required amount without any problems. As a radiation detector in thickness meter used gas-filled argon-xenon ionization chamber (IC) (figure 1).

The main parameters of the IC (figure 2).

– diameter of 165 mm;
– height 80 mm (with electrodes 110 mm);
– case material aluminum thickness 2 mm;
– input window aluminum 0.5 mm thick;
– supply voltage 500 V.

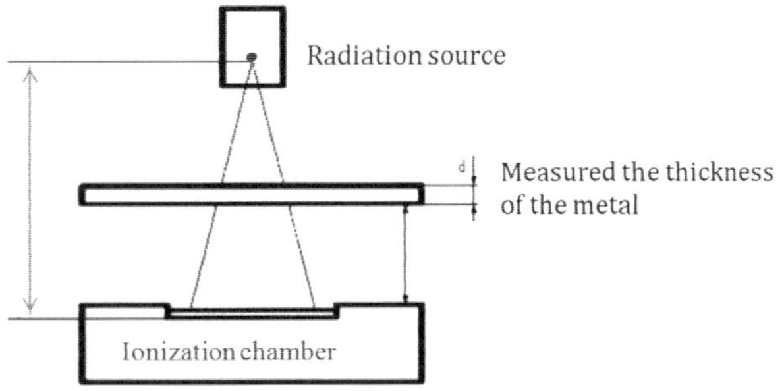

Figure 1. Diagram of the radiation measuring chain gauge FMM-24004.

In the operation of the camera thickness gauges have been exposed to corrosive media and vibration, impact resistance, with time, cameras leak disrupted as a consequence of the pressure of the gas mixture is beginning to subside. Reducing the pressure leads to reduced density and efficiency of the chamber. The number of photons interacted falls, and statistical error increases with a simultaneous decrease in the output level of the camera. Increasing the intensity of the source is not desirable because the lead to overexposure of personnel or require the installation of additional protective shields in the design of devices.

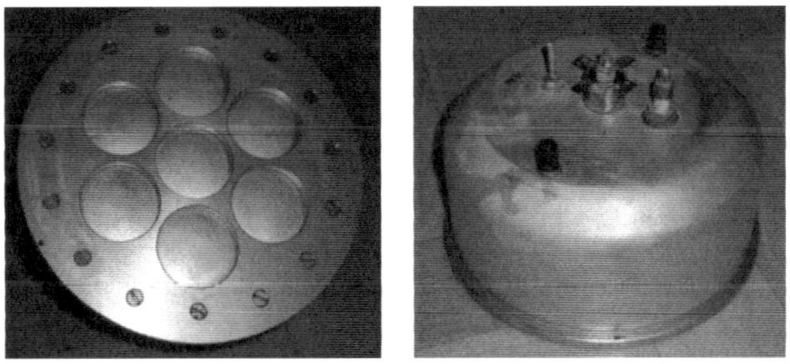

Figure 2. Gas-filled argon-xenon ionization chamber.

Replacing the camera to a more efficient, it is the most simple and secure as well as cost-effective way to solve the problem. As a prototype of heterogeneous camera developed earlier in JSC Scientific-Research Institute of Introscopy "Spectrum" was used for the X-ray thickness gauge RIT10.6 [2–4]. IC design [5] required significant improvement, because effective energy flux in the case of the isotopic source (Americium) instead of the previously used X-ray is increased by 30 %. The dose rate P will be equal to:

$$P = N \cdot E \cdot \mu b \cdot 1/a \ [(\mathrm{R \cdot m^2})/\mathrm{s}] = 7 \cdot 10^4 \cdot 60 \cdot 0.028 \cdot 1/5.5 \cdot 10^{10} = 2.1 \cdot 10^{-6} \ [(\mathrm{R \cdot m^2})/\mathrm{s}],$$

where N – number of photons incident on 1 cm^2 per second at a distance of 1 m from the source; E – energy γ rays in keV; μb – mass emission rate of energy conversion in the air cm^2/g.

When the distance from the source to the center of the chamber 230 mm the dose rate camera geometric center is $40.4\cdot 10^{-6}$ R/s. Current ionization chamber with a volume of 1 cm^3 will make up

$$I_k = 3.3\cdot 10^{-10}\cdot 40.4\cdot 10^{-6} = 1.33\cdot 10^{-14}\ \text{A}.$$

Taking into account the absorption of rolled materials (0.05÷3 mm Fe)

$$I_{k\ min} = I_k\cdot \exp(-\mu\cdot d) = 8\cdot 10^{-16}\ \text{A/cm}^3,$$

where d – maximum thickness of the tape 0.3 cm; μ – linear attenuation coefficient of the material rolled at a photon energy of 60 keV is 9.42 cm^{-1} at its density 7.86 g/cm^3.

Given that modern operational amplifiers don't work well with such small current value, so it is necessary either to increase the source activity, which is unacceptable from the viewpoint of radiation, or substantially no less than $2\cdot 10^4$ to increase the sensitivity of the ionization chamber.

The sensitivity of the ionization chamber can be increased in two ways – increase in the area of the input window and the height of the camera. From the design requirements (TR) the amount of not more than V cm^3 can be increased:

$$V = \pi R^2\cdot h = 3.14\cdot (6.5)^2\cdot 10 = 1.3\cdot 10^3\ \text{cm}^3.$$

Which will provide a current at the minimum and maximum thickness of the rolled:

$$I_{k\ min} = 8\cdot 10^{-16}\cdot 1.3\cdot 10^3 = 10^{-12}\ \text{A};$$

$$I_{k\ max} = 1.33\cdot 10^{-14}\cdot 1.3\cdot 10^3 = 1.7\cdot 10^{-11}\ \text{A}.$$

Current range of the ionization chamber from 10^{-12} to 10^{-11} A insignificant, but to improve the signal/noise ratio is desirable to increase the sensitivity of at least an order of magnitude. For this use as the active material coating the electrode chamber high atomic sputtering with a thickness equal to or less than the mean free path of electrons in the coating material. This sensitivity magnification ionization chamber K_u will be:

$$K_u = \mu_z/\mu_{air} = 4.27/0.0283 = 150,$$

where μ_z – electron photon energy conversion factor for the material chamber electrode with an atomic number Z; μ_{air} – electronic power conversion coefficient of photons in the air.

In view of the not full of electrons from the electrodes, the attenuation of the radiation due to the distance and absorption in the electrode chamber, we assume n = 30 % and Cu = 45 we obtain:

$$I_{k\ min} = 4.5\cdot 10^{-11}\ \text{A (3mm Fe)};\ I_{k\ max} = 7.6\cdot 10^{-10}\ \text{A (0.1 mm Fe)}.$$

To convert the current camera to a digital signal designed electronic circuit measurements with ADC with the integration of the camera signal on the vessel > $50\cdot 10^{-12}$ F. The time constant τ discharge circuit based on its own chamber of the bottle and the ADC input resistance is <50 ms, which is perfectly technical project. Technical demands made by the customer to the ionization chamber and the values obtained from the tests made camera, are shown in table 1.

3. Results and discussion

The camera (figure 3) comprises a body in the form of a hollow metal cylinder with radiolucent conductive caps at the ends, symmetrically placed between the covers along the longitudinal axis of the camera body and parallel plates radiotransparent dielectric material, alternating high and collecting electrodes. Which are formed as coatings of a material with a high Z and insulated from the housing and outer plates forming surfaces rigidly fixed to the housing seals the end planes. The distances between the coatings and high-collecting electrodes respectively and the cavity are sealed, and the chamber is filled with air. Two metal rods, one of which is electrically connected to the high voltage coating electrode, but is insulated from the collecting electrode coatings and other coatings electrically connected to the collecting electrodes, but insulated from high voltage electrode coatings. The rods are arranged parallel to the longitudinal axis of the camera body and rigidly attached to the plates. As a result, high vibration resistance cells were obtained by providing an increased mechanical rigidity of the plates and sensitivity. Reducing inter-electrode distances which increased tension in the chamber at

IOP Conf. Series: Journal of Physics: Conf. Series **808** (2017) 012012 doi:10.1088/1742-6596/808/1/012012

a constant voltage power supply high-voltage electrodes and the same external dimensions of the chamber, which generally improves the dynamic and metrological performance of the camera.

Table 1. The main parameters of the IC.

Parameter	Presentation of the value	Value	Obtained technical requirements
Sensitivity, A·s/Gy	$9.5 \cdot 10^{-5}$	$2 \cdot 10^{-3}$	Yes
Dynamic range, rel. un.	10^3	10^3	Yes
Supply voltage, V	400	150÷500	Yes
Overall dimensions, mm	Diameter 165 Height 125	Diameter 165 Height 125	Yes

Figure 3. Heterogeneous ionization chamber.

4. Conclusions

Depending on the current dose rate were obtained during testing chamber for two focal lengths (figure 4). Thus it is shown that dose rates in the range of 0.1 to 100 P/min, which corresponds to a dynamic range of 10^3, the sensitivity of the camera is constant and equal to $2 \cdot 10^{-3}$ A·s/Gy. Changing the camera voltage range from 150 to 500 insignificantly affects the sensitivity (figure 5).

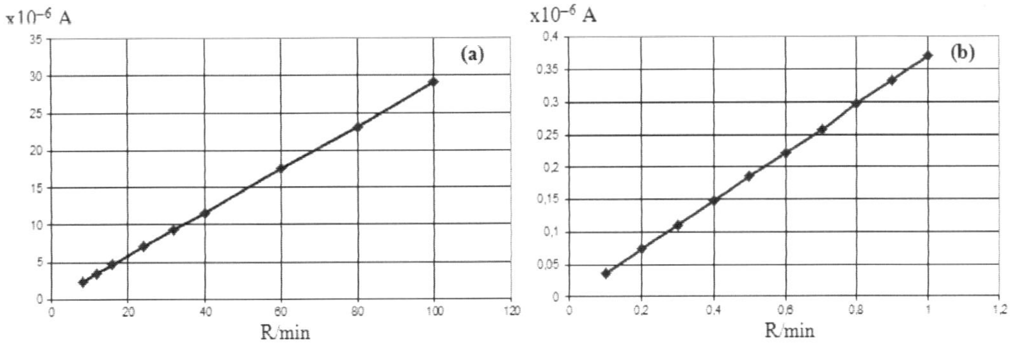

Figure 4. Dependence of the current of the ionization chamber of the dose rate for the focal length of a 0.2 m (**a**) and 2 m (**b**).

Developed by the camera can successfully replace regular detectors on all thickness gauges series FMM-24004, which will increase the metrological characteristics of meters thick and extend their life at work.

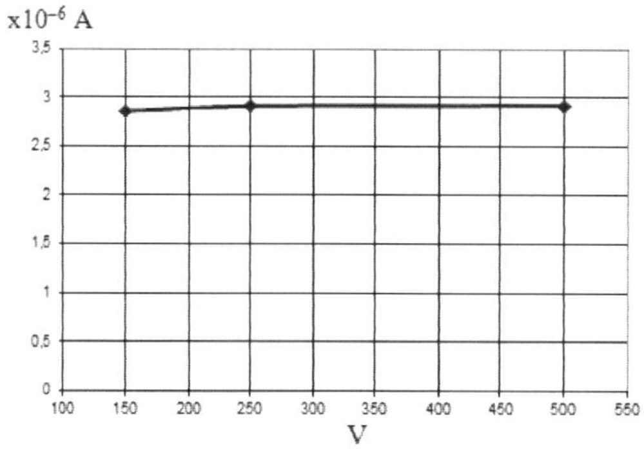

Figure 5. The dependence of the sensitivity of the ionization chamber from the supply voltage.

References

[1] GOST 24034-80. Non-destructive testing radiation. Terms and Definitions
[2] Artemiev B V, Maslov A I, Potapov V N and Vedernikov M B 2003 *Russian journal of nondestructive testing* **6** 55–61
[3] Artemiev B V and Shubochkin A E 2014 *Testing.Diagnostics* **2** 24–31
[4] Klyuev V V, Artemiev B V, Artemyev I B and Klyuyev Z V 2015 *Proceedings of the higher educational institutions. Mechanical Engineering* **7** (664) 67–78
[5] Kanter B M, Vladimirov L V, Vladimirov Yu L and Artemyev I B 2016 Patent RU 163274. Communicating ionization chamber

Small-sized X-ray apparatuses for industrial defectoscopy

E Peliks

LLC "Spectroflash", 194223, Saint Petersburg, Russia

E-mail: info@spectroflash.ru

Abstract. One of the main problems during designing of X-ray apparatuses for the operation in unsteady condition is the problem of designing a unit of minimal weight and dimensions. There are two key priorities: the development of pulsed X-Ray generators on the basis of explosive electron emission and the development of classic thermionic X-ray generators operating at high frequency of transformation of supply mains' voltage to voltage applied to the X-ray tube.

One of the main problems during designing of X-ray apparatuses for the operation in unsteady condition is the problem of designing a unit of minimal weight and dimensions. Although in recent years most manufacturers involved in designing the X-ray equipment have turned to the production of generating monoblocks operating at high frequency of transformation of supply mains' voltage to high voltage, however, the weight of these monoblocks is usually 15–30 kg. This fact can make the use of monoblock X-ray detecting devices under route and mounting conditions very difficult.

For this very reason, the efforts of LLC "Spectroflash" specialists have been always aimed at the development of portable X-ray generators. There were two key priorities:

1. The development of pulsed X-Ray generators on the basis of explosive electron emission;

2. The development of classic thermionic X-ray generators operating at high frequency of transformation of supply mains' voltage to voltage applied to the X-ray tube.

The phenomenon of explosive electronic emission was discovered by a group of Soviet scientists (G A Mesyats, G N Fursey, V A Zuckerman and others) in the late 20th century and since then it has been used by manufacturing of cold-cathode X-ray tubes. The nature of this phenomenon is as follows: at certain values of electric current in the tube diode as a result of warming-up of local apices of the cathode by own self-emission current, they explode and dense plasma forms, that is an intense source of electrons.

Many experimental studies showed that in the initial period of time that equals to several nanoseconds, the velocity of plasma expansion into a vacuum for most metals is $(1–3)\cdot 10^6$ cm/s, so it shall be assumed that current in this tube diode is limited by a space charge in the gap "the front of moving plasma–anode". That means that such a diode is the tube diode with a decreasing "anode–cathode" gap. Appeared cathode plasma leads to the formation at the cathode of the tube diode a liquid constituent of metal from which under the influence of electric field micro-apices are drawed and craters appear at the solid part of the cathode. In such a manner, a specific cathode micro-relief is created. With each current pulse, a certain quantity of cathode micro-apices disappears and a certain quantity of them appears again. This process is called the effect of a self-healing cathode. Explosive electronic emission is quite a complicated phenomenon and that's why it is not a subject of a detailed description in the current report.

From all has been said above, only several practical assumptions can be made:

1. The plasma cathode under certain conditions can be considered as a regular metal cathode, so at determining of the X-ray tubes' voltage–current characteristics with explosive electronic emission, it is possible to use all instruments developed for regular diodes with an incandescent cathode, taking into account that current in this case increases in comparison with Child–Langmuir equation (three-halves power law) due to the decrease of the "anode–cathode" gap by the plasma movement;

2. There is certain dependence between electric field intensity at the tube cathode, the current amplitude and the tube life duration. This refers to the realization of the mode of the "self-healing cathode".

These assumptions define the list of requirements to output parameters of a high-voltage source, feeding the X-ray tube with an explosive cathode.

First of all, the source has to produce at the output a high-voltage pulse with amplitude that provides electric field intensity at the tube cathode about 10^6 V/cm. At that, pulse front time shall not exceed units of nanoseconds. Only under these conditions the process of explosive electronic emission is developed on the tube.

And, in the second place, the current pulse, appearing in a tube vacuum gap, on the one hand, has to be powerful enough to obtain maximal energy of the X-ray emission, and, on the other hand, its duration has to be less than the time required for plasma to cut the "anode–cathode of the tube" gap. Elsewise, there will be an electric arc in the tube, leading to a quick breakdown of the tube.

In all pulse apparatuses produced by LLC "Spectroflash", Tesla transformer as a high-voltage source is used. Due to a special construction of the transformer, the mutual-coupling coefficient between its coils is 0.65–0.7, which provides a high performance coefficient of the whole generator. The voltage pulse duration at the output of the generator is 1–2 ms depending on the type of apparatus. To obtain the duration of pulse at the X-ray tube about 1–1.5 ns between the generator and the X-ray, a special discharger-sharpener filled with hydrogen under pressure of 40–60 atm. In case of electrical breakdown of this discharger, a high-voltage pulse with front less than 1 ns on electrodes of the X-ray tube appears. With this front the explosion of cathode micro-apices with further plasma formation happens. All three elements (Tesla transformer, discharger-sharpener and X-ray tube) at such short voltage pulses have extremely low weight and dimensions.

Pulse X-ray apparatuses have been produced by LLC "Spectroflash" company for 25 years. Its first series contained 5 models and had a common name ARINA. These apparatuses are still produced. Their technical characteristics are shown in the table 1.

Table 1. Technical characteristics of ARINA pulse X-ray apparatuses.

Characteristics	ARINA-1	ARINA-3	ARINA-7	ARINA-9
X-ray tube operating voltage, kV	150	200	250	300
X-ray dose at a distance of 500 mm from focus of X-ray tube during 1.5 min, mR	500 (in 1000 p.)	600	1000	1200
Focal spot size (mm)	2.5	2.5	2.5	2.5
Ray method – directed/panoramic. Maximal thickness of X-ray steel from focal distance 700 mm during exposure period of 10 minutes with film usage:				
D7 + Pb	–	–	20	25
F8 + RCF	15	20	25	30
F8 + NDT 1200	25	30	40	50
Weight of X-ray source, kg	8.9	5.5	7.9	8.1
Dimensions of X-ray source	530×125 ×210	420×125 ×215	515×140 ×210	510×135 ×215

At the present time, a new series of pulse apparatuses named PAMIR is developed. As distinguished from the ARINA apparatus, new detecting devices have increased operational life and longer time of continuous operation. X-ray characteristics of both series are the same. The weight and dimensions of apparatuses of both series are close to each other. Pulse X-ray apparatuses are shown in figure 1.

Figure 1. Pulse X-ray apparatuses (from left to right) PAMIR-200, PAMIR-250, PAMIR-300.

A thermionic group of apparatuses is produced under MART name and consists of 2 models. Technical characteristics of both apparatuses are shown in the table 2.

Table 2. Technical characteristics of MART thermionic X-ray apparatuses.

Characteristics	MART-200	MART-250
X-ray tube operating voltage, kV	205–200	130–250
X-ray dose at a distance of 500 mm from focus of X-ray tube during 1 min, R	3	5
Max anode power, W	100	200
Ray method – directed/panoramic. Maximal thickness of X-ray steel from focal distance 700 mm during exposure period of 10 minutes with film usage:		
D7 + Pb	20	30
F8 + RCF	30	40
F8 + NDT 1200	40	50
Power supply	220 V 50 Hz	220 V 50Hz
Weight of X-ray source, kg	6	9
Dimensions of X-ray source	430×100×160	580×140×190

MART differential characteristics are: first of all – high transformation frequency of 220 V supply mains voltage to 200–250 kV voltage, required for supply to the X-ray tube, and secondly – bi-polar connection of high-voltage to X-ray tube electrodes. Both factors allowed decreasing significantly the weight and dimensions of X-ray generators. Practically, their weight is commensurate with the weight of X-ray generators but both MART apparatuses are significantly more powerful than pulse apparatuses. Each of them provides for the possibility of adjusting the anode voltage in wide range and the total operational life of each apparatus exceeds the operational life of pulse apparatuses. In both apparatuses, 0.1 BPM27 X-ray tube, with two versions, is used. The MART-250 apparatus' tube has a side output in the solid angle 50°. In the MART-200 there is a front output in the solid angle 140°.

Figure 2. Thermionic group of apparatuses MART-250, MART-200.

Summarizing all above-mentioned information, it is possible to make the following general conclusions:

1. Pulse and high-frequency apparatuses are close to each other in their dimensions and weights and X-ray characteristics, so they can be used in the same conditions;
2. Pulse apparatuses can be used both for directed and panoramic X-raying, they have extremely simple construction, and they are repairable at a customer's site, comparatively cheap in operation, undemanding in all weather conditions;
3. High-frequency apparatuses have the possibility of adjusting the X-ray tube voltage; they are more powerful in comparison with pulse apparatuses and have longer operational life.

References

[1] Kluev V V, Sosnin F R, Filinov V N et al. 1995 *Non-destructive testing and diagnostics. Reference book* (Moscow: Maschinostroenie)
[2] Mesyats G A, Ivanov S A, Komyak N I and Peliks E A 1983 *Powerful nanosecond pulses of the X-ray emission* (Moscow: Energoizdat)
[3] Merkulov B P, Peliks EA and Samorodov V G 2003 Patent RU 2251230. Pulsed X-ray generator
[4] Peliks E A 2007 *Proceedings of the 9th Conference on non-destructive testing (Kiev)*
[5] Peliks E A 2006 *Proceedings of the XIV Annual International Conference "Modern methods and means of non-destructive testing and technical diagnostics" (Yalta)*

IOP Conf. Series: Journal of Physics: Conf. Series **808** (2017) 012014 doi:10.1088/1742-6596/808/1/012014

Portable technical equipment for X-ray diagnostics in unspecialized conditions

N N Potrakhov[1], Y N Potrakhov[1], A Y Vasilev[2] and V A Lyubimenko[3]

[1] Department of electronic instruments and devices, Saint Petersburg Electrotechnical University "LETI", 197376, Saint Petersburg, Russia
[2] Moscow State University of Medicine and Dentistry, 127473, Moscow, Russia
[3] Saint Petersburg City Children's Hospital № 1, 198095, Saint Petersburg, Russia

E-mail: yus87@yandex.ru

Abstract. This article presents the results of research in the field of creation of portable X-ray diagnostic apparatuses for medical purposes. Preconditions for X-ray studies in nonstationary and nonspecialized conditions are provided. The design and examples of the use of portable X-ray diagnostic complex for neonatology are described.

1. Introduction

Currently, all the more urgent becomes the problem of X-ray studies in unspecialized conditions, such as patient "at home". Reasons for conducting X-ray diagnostic studies at home may be different, including:

– somatic "serious" condition in elderly patients who are unable to get to the health facility for suspected pneumonia, fractured limbs, other trauma, etc.;

– availability of financial well-off citizens, which are ready to pay for X-ray diagnostic "at home".

However, in many cases, X-ray radiography in unspecialized conditions cannot be regarded as a method of definitive diagnosis, since there are a number of specific restrictions in carrying out such researches [1, 2]. At first, often medical condition of patient does not allow performing X-ray examination in the standard (at least 2) projections; secondly, the characteristics of the technical means of X-ray radiology do not allow conducting the X-ray studies.

Obviously, the use of traditional stationary tripod devices providing positioning X-ray apparatus relative to the patient, at home is not possible. Therefore, staff use portable tripods or takes the picture without using a tripod – "in hand." Therefore, special importance becomes the issue of radiation safety for the staff carrying out the study, and to the others who may also be involved in the study, such as the laying of the patient. However, the use of stationary means of protection against the unused X-ray radiation, as well as remove others at a safe distance in the home is almost impossible. Therefore it is necessary to use such a technique of X-ray photography, which can significantly reduce the exposure dose of radiation than when using stationary devices. In this case, is extremely important to increase the image quality of the test authority, the necessary and sufficient for a decision of the presence or absence of pathology.

Of particular importance to the issue of X-ray diagnostics in unspecialized conditions gives the important challenges of modern clinic – radiological surveys in neonatology. Additional difficulties in its decision are caused by the small size and weight of the newborn (up 400 g); conditions in which

IOP Conf. Series: Journal of Physics: Conf. Series **808** (2017) 012014 doi:10.1088/1742-6596/808/1/012014

there is a child (incubator, open resuscitation system, operating), as well as the severity of his condition (reanimation, intensive care, rehabilitation). Often, because of the condition of the patient: artificial ventilation, cardiac or infusion therapy, active aspiration – X-ray study can be conduct only in the reanimation.

2. Results and discussion

According to City Children's Hospital № 1 (Saint Petersburg) only in its resuscitation units annually around 4500 radiographic procedures in 2 or 3 projections. In total, it is up to 10 thousand X-rays in a year. Currently, these studies are carried out on the domestic or foreign X-ray machines, practically not adapted to the problems and conditions of the resuscitation of newborns with pathologies. Accordingly, the additional radiation load subjected to medical staff and other patients [3].

Research carried out in collaboration with experts from Federal Almazov North-West Medical Research Centre and City Children's Hospital № 1 showed that virtually all requirements to the organization of the X-ray studies in unspecialized environment and, above all, neonatology, satisfied by method of microfocus X-ray radiography [4].

Its main features are:

– using X-ray apparatus with a size of focal spot of X-ray tube less than 0.1 mm;

– 1.5–2 fold increase voltage on X-ray tube and 3–5 times decrease focal distance as compared to standard X-ray technique.

To implement a methodology was developed specialized digital X-ray diagnostic complex "PARDUS-Neo" composed of the X-ray apparatus "PARDUS" in a portable version [5] and X-ray visualisation system on the basis of the screen with the photoluminophore. The construction of the first domestic portable apparatus "PARDUS-R" allows obtaining X-ray images without the use of a stationary tripod. X-ray lab assistant takes a picture while holding the device in hand, while being in the radiation safety conditions (figure 1).

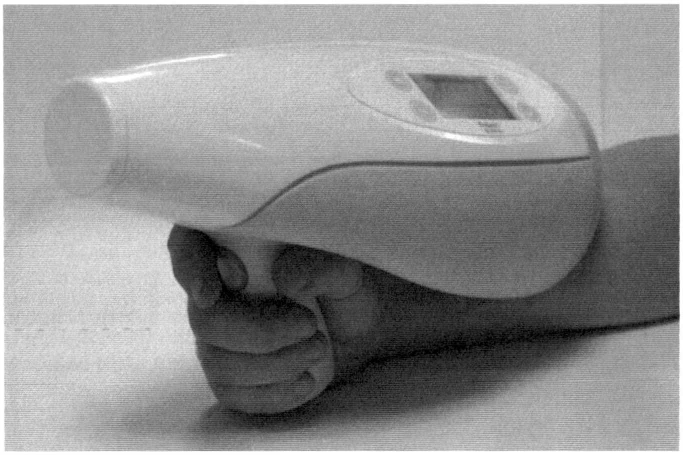

Figure 1. Portable X-ray apparatus «PARDUS-R».

Figure 2 shows an X-ray image of the preparation of the newborn body obtained by the method of microfocus X-ray radiography in unspecialized conditions on the X-ray diagnostic complex "PARDUS-Neo". On the X-ray image is good visualized structure of bone elements: bone of the skull base, thorax skeleton, spine, pelvis and long bones, and soft tissues, including non-contrast nasotracheal tube.

Figure 3 shows the X-ray images of the thoracic department (patient P., age 7 months, weight 3.5 kg), obtained in a stationary X-ray room in City Children's Hospital № 1 (figure 3(a)), and obtained in unspecialized conditions "at home" for the same patient (figure 3(b)) using the complex "PARDUS-Neo".

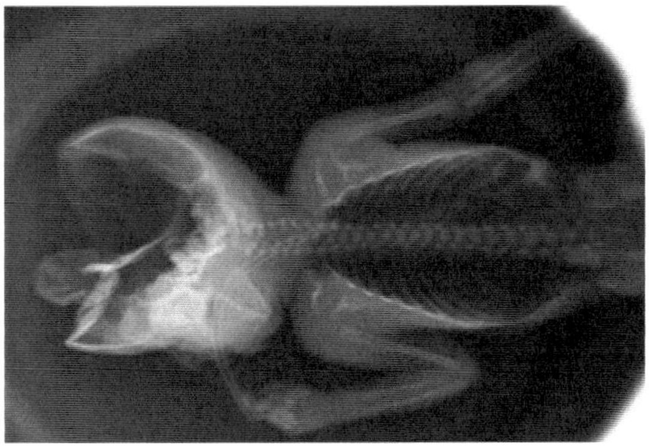

Figure 2. X-ray image of the preparation of a newborn body.

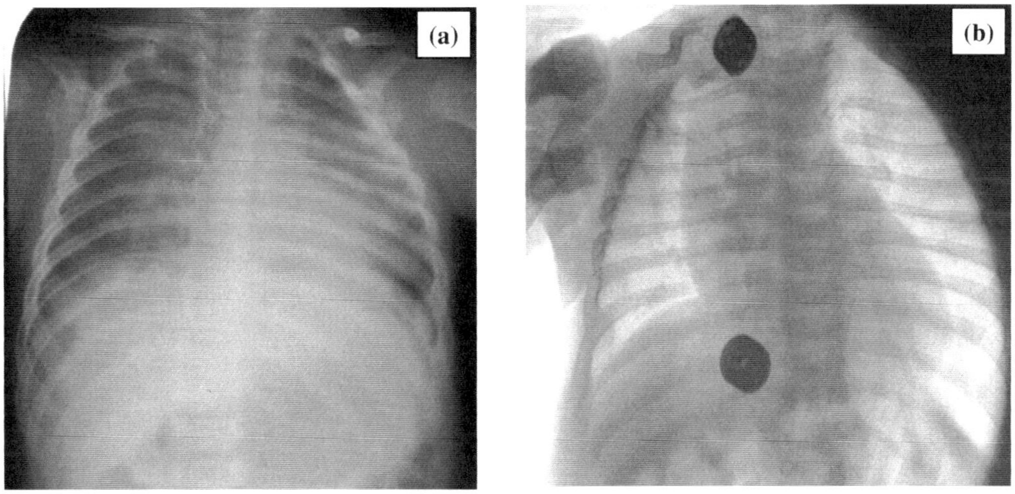

Figure 3. X-ray images of the thoracic department: (**a**) – stationary X-ray room in City Children's Hospital № 1; (**b**) – unspecialized conditions "at home" using the complex «PARDUS-Neo».

3. Conclusion

According to experts the informativeness of the X-ray image obtained at home, is almost as good as X-ray image obtained in the hospital. Thus, the described results of research open the prospect for the use of techniques of microfocus X-ray portable technical equipment for newborns research.

References

[1] Potrakhov N N, Potrakhov E N, Gryaznov A Y, Vasilyev A Y, Balitsky N V and Boychak D V 2013 *Biomedical Engineering* **5** 183–5
[2] Potrakhov N N 2015 *Biomedical Engineering* **5** 237–40
[3] Potrakhov N N, Potrakhov E N, Gryaznov A Yu and Vorob'ev B F 2008 *Biomedical Engineering* **6** 312–3
[4] Potrakhov N N and Mazurov A I 2005 *Biomedical Engineering* **2** 77–8
[5] Gryaznov A Yu, Potrakhov E N and Potrakhov E N 2008 *Biomedical Engineering* **5** 270–1